W0036500

A Journey into Modern Physics

Carmine Granata

A Journey into Modern Physics

From Relativity to Quantum Technologies

 Springer

Carmine Granata
Institute of Applied Sciences and Intelligent Systems
National Research Council (CNR)
Pozzuoli, Napoli, Italy

ISBN 978-3-031-77774-5   ISBN 978-3-031-77775-2   (eBook)
https://doi.org/10.1007/978-3-031-77775-2

© The Editor(s) (if applicable) and The Author(s), under exclusive license to Springer Nature Switzerland AG 2025
This work is subject to copyright. All rights are solely and exclusively licensed by the Publisher, whether the whole or part of the material is concerned, specifically the rights of translation, reprinting, reuse of illustrations, recitation, broadcasting, reproduction on microfilms or in any other physical way, and transmission or information storage and retrieval, electronic adaptation, computer software, or by similar or dissimilar methodology now known or hereafter developed.
The use of general descriptive names, registered names, trademarks, service marks, etc. in this publication does not imply, even in the absence of a specific statement, that such names are exempt from the relevant protective laws and regulations and therefore free for general use.
The publisher, the authors and the editors are safe to assume that the advice and information in this book are believed to be true and accurate at the date of publication. Neither the publisher nor the authors or the editors give a warranty, expressed or implied, with respect to the material contained herein or for any errors or omissions that may have been made. The publisher remains neutral with regard to jurisdictional claims in published maps and institutional affiliations.

This Springer imprint is published by the registered company Springer Nature Switzerland AG
The registered company address is: Gewerbestrasse 11, 6330 Cham, Switzerland

If disposing of this product, please recycle the paper.

To my father

Preface

Quantum physics and Albert Einstein's theory of Relativity have completely changed the way we look at the world by introducing completely counter-intuitive concepts that are in stark contrast with the common sense. The aforementioned theories, also referred to as *modern physics*, have also allowed a real technological revolution, penetrating deeply into everyone's life.

The purpose of this book is to provide an overview of the foundations and principles on which modern physics is based, as well as its major applications, whose technological impact has been truly remarkable. It also aims to mention that branch of contemporary physics known as the *second quantum revolution* and its current and future applications, referred to as *quantum technologies*.

In order to better understand the spirit in which this book was written, it is appropriate to reflect on scientific communication and dissemination.

In modern society, knowledge in all areas is growing at an ever-increasing pace, and technological progress is advancing in step. On the other hand, it is becoming increasingly difficult to communicate and disseminate scientific results, leading to a consequent distrust of science by people, but the importance of communicating and disseminating the progress of science to the general public is undeniable. That said, it is equally clear that the way in which a scientific topic is disseminated is of fundamental importance. Often, there is a tendency to be very technical, almost immediately alienating the interest of people who do not have that technical knowledge, while at other times, the opposite extreme is reached, risking completely distorting and/or mystifying the concepts. Perhaps a middle ground might be the best approach, that is, while maintaining a language understandable to many, try not to lose the scientific rigor necessary not to alter the content of the topic. It is clear that this is not easy and that behind good dissemination, whether on television or in magazines and books, there is a lot of work and talent.

In the particular case of modern physics (theory of Relativity and Quantum mechanics) there are additional difficulties related to the impossibility, with a non-specialist audience, of using advanced mathematical language. To this is added the need to expose completely counter-intuitive concepts and outside of common sense, typical of modern physics, and in some cases the possible metaphors are only misleading. The same teaching of these topics at the university level does not have a consolidated teaching protocol, as is the case for classical physics. In the particular case of quantum mechanics, sometimes a chronological historical approach to the topics is preferred, which however foresees a particularly difficult beginning (black body spectrum), in other cases an axiomatic setting very similar to a pure mathematical theory is preferred. Personally, I believe that both at the dissemination level and in teaching, the historical approach is more effective as it helps to create an apparent logical thread that aids understanding. It is in this perspective that this book was written.

The first chapter is dedicated to one of the most beautiful theories of physics, Albert Einstein's Special and General Relativity, highlighting the implications that the theory has also had in astrophysics and cosmology (black holes, expansion of the Universe, gravitational waves).

The second chapter illustrates the foundations and principles of quantum mechanics, starting from the first ad hoc hypotheses that contemplated the quantum behavior of the nature to explain some phenomena inexplicable with the physical knowledge of the late nineteenth century. We then arrive at the formulation of quantum theory and its fundamental equation (Schrödinger's equation). The foundations of the relativistic version of quantum mechanics and its implications in the world of the infinitely small and fundamental forces are then discussed.

In the third chapter, we move from the infinitely small to the macroscopic world, highlighting how quantum mechanics is able to explain phenomena that concern condensed matter. The same chapter provides an overview of the major and most important applications of quantum mechanics, which shows the enormous impact of this bizarre theory in everyday life.

In the last chapter, the conceptual foundations and paradoxes of quantum mechanics that were at the basis of the famous controversies between Albert Einstein and Niels Bohr are exposed, but at the same time inspired theories and extraordinary experiments, leading to the development of new quantum technologies that will most likely enter everyday life in the near future, just like lasers, electronics, LED bulbs, and nuclear medicine. Some final paragraphs are dedicated to these new quantum technologies, illustrating the potential of quantum computing, quantum cryptography and teleportation, and other quantum technologies.

Unlike most popular physics books, in this book, a reasonable space has been allocated to the applications of quantum physics in order to convey the reader not only its conceptual importance but also its applicative one. The downside of this choice is the risk of being a bit technical, but the advantage is the awareness that the reader does not ask the fateful question: but what is all this for in practice?

Another important difference is the space dedicated to the physics of matter which, in addition to being the basis of most applications of quantum physics, represents a very fascinating topic that is usually overlooked in popular physics books.

Thanks to the aforementioned aspects and the last chapter on quantum technologies, this volume provides a complete overview of modern physics, emphasizing routine applications and more futuristic ones linked to new quantum technologies.

The reader will have the opportunity to encounter the most important topics of modern physics: from the theory of Relativity with its countless implications, to quantum physics with its applications to the physics of elementary particles and condensed matter physics up to quantum computing. All this is described by making numerous references to important and fundamental discoveries, both historical and contemporary. Therefore, updates and current advancements in research in many of the fields covered are also provided.

The illustrations, many of which are originally elaborated by the author, help to understand the concepts of physics as well as the operating principles of the various applications reported.

The book is essentially popular, without using complicated formulas or particular technicalities; therefore, it does not require in-depth knowledge of physics or mathematics; the knowledge acquired in high schools is sufficient to understand the topics covered. A systematic reading from the beginning is recommended, as some concepts and definitions are then subsequently recalled.

Before embarking on our journey into the extraordinary world of modern physics, I would like to remind you that 2023 coincided with the centenary of the foundation of the National Research Council (CNR), the largest public research body in Italy. With its seven departments, 88 research institutes and about 8500 employees among researchers, technologists, technicians, and administrative staff, the CNR carries out advanced research in almost all scientific sectors contributing to the growth of one of humanity's most precious assets, *knowledge*. Among its presidents, we remember figures of very high scientific stature such as its founder and first president Vito Volterra (mathematician, physicist and politician) and the great inventor and politician Guglielmo Marconi, Nobel laureate in physics in 1909 for the development of wireless telegraphy.

Pozzuoli, Italy   Carmine Granata

Acknowledgments

I wish to express my thanks to my friends and colleagues Francesca Alesse (Philosopher, CNR Research Director) and Carmela Bonavolontà (Physicist, CNR Researcher) for their valuable suggestions to make the text clearer and more readable.

Contents

1

Einstein's Theory of Relativity: A New Vision of the World

In September 1900, one of the most authoritative physicists of the time, William Thomson, better known as Lord Kelvin, at the assembly of the British Association for the Advancement of Science, in Bradford, pronounced the following phrase: "There is nothing new to discover in physics anymore. All that remains to be done are ever more precise measurements". And it is precisely from some of the measurements to which Kelvin was referring that the theories of Relativity and quantum physics were born, which together with the infinitesimal calculus developed by Isaac Newton and Gottfried Leibniz about two centuries earlier, can be considered among the highest peaks conquered by human intellect. This chapter is dedicated to Einstein's theory of Relativity, considered one of the greatest masterpieces of physics.

1.1 The Theory of Special Relativity: A New Conception of Time and Space

Einstein's theory of Special or Restricted Relativity is a generalization of the laws of mechanics developed a few centuries earlier by Galileo Galilei and Isaac Newton with particular reference to systems in relative motion with constant speed and implies a new and revolutionary concept of space and time. The theory manifests its effects for very high speeds or non-negligible compared to the speed of light (about 300,000 km/s) and reduces to the classical theory for speeds much smaller than the speed of light. In this latter regime, relative motions are well described by the laws of classical Relativity

© The Author(s), under exclusive license to Springer Nature Switzerland AG 2025
C. Granata, *A Journey into Modern Physics*, https://doi.org/10.1007/978-3-031-77775-2_1

developed by Galileo in his famous book "Dialogue on the Two Chief World Systems", a book for which the Paduan scientist, in 1633, was accused by the Inquisition court of heresy for his manifest positions in favor of the heliocentric Copernican system.

The principle of Galilean Relativity maintains that the laws of mechanics must have the same form in any inertial reference system, i.e., a system that moves at a constant speed. A reference system is a system with respect to which a certain physical phenomenon or an object is observed and measured. Examples of reference systems can be the Sun, the Earth, a room in our apartment, a ship, a train, a car, or even a body. To explain the principle of Relativity, Galileo gives the example of a ship traveling on a calm and flat sea and maintains that any observation or experiment an observer makes in the ship's hold obtains exactly the same results obtained by an experimenter stationary at the port. This is actually a principle that we experience almost every day: if we travel on a train or in a car that moves along a straight path at a constant speed and we bounce a coin in our hand or throw something inside the train or car, we observe the same behavior that we would have observed from a room in our house. Or, to use an example similar to that of Galileo, the flight of a fly trapped in our car is exactly the same as what we observe when we are stationary. Therefore, the laws of mechanics and the related equations must have the same form, they must be *invariant* when moving from one reference system (train, ship) to another (station, port). But to move from one reference system to another, transformations of speeds and spatial coordinates must be made, which in the case of inertial systems are simple and easily intuitive. Let's consider a simple case: if a man on a train throws a ball in a horizontal direction at a speed of 40 km/h and the train moves in the same direction and in the same sense at a speed of 30 km/h, for a stationary observer at the train station the speed of the ball will be given by the sum of the speed of the ball relative to the train and the speed of the train, i.e., 70 km/h. Therefore, the transformation of the speed will be $v_p = v_T + v'_p$ where v_p is the speed of the ball relative to the station, v_T is the speed of the train and v'_p is the speed of the ball relative to the train. If we consider two systems of Cartesian axes (Fig. 1.1), one fixed with the train and the other with the station, it is easy to find the law of transformation of the spatial coordinates. Simply the distance traveled by the ball at a certain time t (coordinate x) in the system fixed to the station will be given by the sum of the distance traveled by the ball relative to the train (coordinate x′) and the distance traveled by the train relative to the station which is equal to the speed of the train multiplied by the time *t*. So, in the end we will have: $x = x' + v_T \cdot t$. This transformation of speeds and spatial coordinates (*Galilean transformations*), so obvious and reasonable in the case

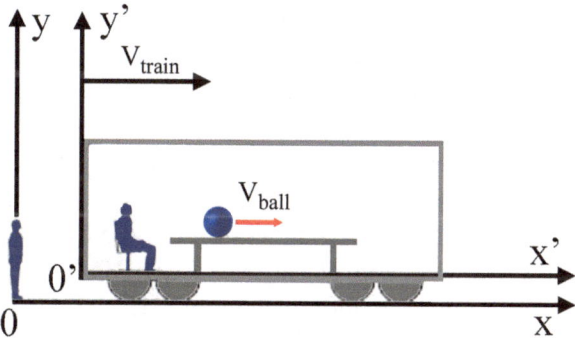

Fig. 1.1 Schematic representation of two systems of reference in relative motion. For the stationary observer, the speed of the ball will be equal to the sum of the speed of the train and that of the ball relative to the train. In the theory of Special Relativity, this obvious law of composition of speeds is not valid

where the speed of the train was very high or, more precisely, not negligible compared to the speed of light no longer apply and go, as we will see later, replaced with more general transformations that reduce to those of Galileo when the speeds involved are small compared to that of light. In summary, according to Galilean Relativity the laws of mechanics are invariant for Galilean transformations. In physics when these conditions occur, we also speak of symmetry with respect to certain transformations.

It is unlikely that Einstein decided out of the blue to develop a new theory of mechanics without having valid assumptions, also because Galilean and Newtonian mechanics was well established and has allowed over the years to make very accurate predictions as well as applications of great importance such as ballistics, celestial mechanics or astrodynamics. Currently the design of space flights, the guidance of space probes, the calculation of aircraft routes are based on the laws of Newtonian dynamics. Usually when the foundations of a consolidated theory are questioned, we start from current problems of apparently different nature whose resolution leads to review and extend an existing theory. Therefore, to understand the genesis of the theory of Relativity, it is appropriate to make a brief state of the art of physics at the end of the nineteenth century.

The nineteenth century was certainly characterized by the study of electrical and magnetic phenomena. In 1785, Charles Coulomb formulated the law according to which two electric charges attract each other if they have opposite signs, or repel each other if they have the same sign, with a force that is directly proportional to the product of the charges and inversely proportional to the square of their distance. In 1820, Hans Ørsted discovered that the magnetic needle of the compass changes direction when it is placed near a wire carrying an electric current, highlighting the first connections between electrical and magnetic phenomena. In 1831, the British physicist Michael Faraday and, independently, the American physicist Joseph Henry discovered that a sudden displacement of a magnet produces an electric current in a conductive wire present in the immediate vicinity and vice versa: if the same wire is in motion in the proximity of a magnet, it is crossed by an electric current. They discover the phenomenon of electromagnetic induction which is the basis of power stations, electric motors, and electric transformers. A groundbreaking discovery that has completely changed our way of living.

In 1864, the Scottish physicist James Clerk Maxwell presented to the Royal Society of London the famous four equations, known as Maxwell's equations, which excellently synthesize and formalize the theory of electrical and magnetic phenomena and predicted the existence of electromagnetic waves propagating at a constant speed, namely the speed of light. Following the experimental verification of the existence of electromagnetic waves by Frederick Hertz in 1886, it seemed natural to the scientific community to hypothesize a medium (the *ether*) through which the new waves would propagate and since the speed that appeared in Maxwell's equations was a constant, it was hypothesized that the reference system associated with the ether was an absolute and immobile system.

There was also another very important point: when a Galilean transformation was performed, that is, when one moved from one inertial reference system to another, the Maxwell's equations changed their form, they were therefore not invariant to Galilean transformations.

The hypothesis most credited by the scientific community of the time was the following: the principle of Galilean Relativity is valid only for mechanics and electromagnetic phenomena can only be described in a privileged reference, that of the ether. The other possibility, contemplated by few scientists including Einstein, foresaw the validity of the principle of Relativity also for electromagnetic phenomena, but the Galilean transformations were not correct and had to be replaced by other more general transformations.

At this point, an experiment was needed to shed light on these fundamental doubts. In 1887, in one of the most important experiments in physics, two American experimental physicists, Albert Michelson and Edward Morley, demonstrated the non-existence of the ether, a medium through which electromagnetic waves, and therefore also light, should have propagated, laying the experimental foundations of special or restricted Relativity. Therefore, the famous experiment of Michelson and Moreley threw into crisis the hypothesis of the ether, widely spread and accepted in the scientific community of the late nineteenth and early twentieth centuries. Even today, referring to television broadcasts, the term *via ether* is sometimes used.

Here comes into play an unknown young physicist employed at the patent office in Bern, named Albert Einstein who in 1905 wrote three scientific articles destined to become milestones of modern physics. In one of these "On the Electrodynamics of Moving Bodies", Einstein presents his theory of special or restricted Relativity. In the introduction, Einstein immediately reveals his strong opposition to the existence of the ether and of an absolute reference system: "The introduction of a *luminous ether* has so far proved superfluous, as according to the interpretation developed, no *absolute space at rest* with special properties is introduced…". Moreover, he underlines how the interpretation of electromagnetism, especially when considering moving bodies (hence the title of the famous article), leads to asymmetries that he could hardly tolerate. In particular, there are some electromagnetic phenomena, such as the interaction between a magnet and a conductor, which according to Einstein depend only on their relative motion, instead depending on whether the magnet or the conductor is in motion or still, they are considered distinct phenomena according to the interpretation of classical physics.

Einstein, starts from two hypotheses that he elevates to principles or postulates:

1. All the laws of physics including those of electromagnetism must preserve their form when moving from a reference system that moves in a straight line at a constant speed (inertial system) to another;
2. The speed of light is always the same regardless of the source that emitted it. Moreover, the speed of light is a speed limit: nothing can travel faster than light.

The first principle of Einstein's theory of Relativity seems very similar to Galileo's, but in reality, in addition to considering all the laws of physics including those related to electromagnetic phenomena, it assumes a more general and less empirical character. Einstein considered it to be a principle of symmetry of fundamental importance. In this case, the symmetry is not with respect to Galileo's transformations, but to other more general transformations. For over 30 years and until his death, Einstein tried to find a single theory that would unify all the existing forces in nature; in particular, his goal was to formulate a theory that would unify quantum mechanics and General Relativity. About 70 years after his death, this titanic task has not yet been accomplished. In short, he was a great unifier, and as such he considered principles of symmetry to be of fundamental importance. In fact, he was not wrong, as symmetries will play a fundamental role in the development of theories to explain the world of elementary particles and fundamental interactions.

The second principle is in clear contrast with Galileo's transformations and with common sense: if a beam of light is emitted from a rocket travelling towards the Earth at a certain speed v, the speed relative to a stationary observer on the Earth is not equal to the sum of the speed of light plus that of the rocket, as Galileo's transformations and everyday experience would seem to suggest, but is always equal to the speed of light c. But if this is true, and it is, then the concept of time must be revised and Galileo's transformations must be replaced by more general transformations that no longer provide an absolute time for all reference systems, but time depends on the reference system in which it is measured.

The concept of time has always been the subject of deep philosophical reflections, as it is difficult to give a comprehensive definition. The famous aphorism of Saint Augustine in his most famous work "The Confessions" well illustrates the idea just stated: "What is, then, time? If no one asks me, I know; if I should explain it to who asks me, I do not know". However, no one before Einstein had questioned the absolute nature of time. It is natural to think that two synchronized clocks will always indicate the same time regardless of their state of motion and the observer. Instead, according to the theory of Relativity, time flows differently depending on the reference system in which we find ourselves, introducing a characteristic of time completely counter-intuitive and outside common sense.

Using these two principles, Einstein arrives at new transformations of space-time variables in which time is no longer absolute but also depends on space and especially on the speed at which the two inertial reference systems move. The concept of space thus merges with that of time and we speak of four-dimensional space-time in which the temporal dimension (t) is added to the three spatial dimensions (x, y, z).

The mathematical formalization of the space-time continuum was developed by the German mathematician Hermann Minkowski, Einstein's mathematics professor at the Zurich Polytechnic. During the meeting of German natural scientists and doctors (September 21, 1908), Minkowski gave a lecture on this topic whose prologue read: "The concepts of space and time that I wish to present to you originate from the field of experimental physics, and therein lies their strength. From now on space by itself, and time by itself, are doomed to fade away into mere shadows, and only a kind of union of the two will preserve an independent reality". The conference summarized the results published by Minkowski in an article of the same year, in which he reformulated Einstein's 1905 work introducing a non-Euclidean four-dimensional geometry, defining a space-time in which the three spatial coordinates and the temporal one were equivalent (Minkowski *space-time*), and to which Einstein initially did not pay much attention, saying: "Since the mathematicians have invaded the theory of Relativity, I no longer understand it myself". Later, however, when he found himself facing the mathematical problems of General Relativity, he recognized the indispensability of Minkowski's four-dimensional scheme. In fact, in the introduction of the article on General Relativity of 1916, Einstein reports: "The generalization of the theory of Relativity is greatly facilitated by the form that has been given to the theory of Special Relativity by Minkowski, the mathematician who first clearly recognized the formal equivalence of spatial and temporal coordinates, and made it usable for the construction of the theory".

Returning to the transformations between two inertial reference systems (known as Lorentz transformations), they were introduced ad hoc in 1904 by the Dutch physicist Hendrik Antoon Lorentz as a mathematical device, with the intent of making Maxwell's equations invariant and to explain the null experiment of Michelson and Morley. Lorentz believed in the existence of the ether, so even if the transformations were correct, the interpretation was not. Instead, Einstein derived them from the two principles of Relativity.

A direct consequence of the Lorentz transformations is the time dilation and space contraction: if a clock in a reference system is observed to be moving uniformly in a straight line, it appears slower than a clock attached (stationary) to the reference system, just as, if an object in motion is observed, it appears shorter than the same stationary object in the reference system. From the Lorentz transformations, the laws of transformation of velocities and accelerations between two reference systems in relative motion to each other at constant speed can be derived. Compared to those of Galileo, these are more complex transformations in which the speeds do not simply add up, as done in the example shown in Fig. 1.1, since the speed of light is an insurmountable limit.

Although the procedure used to formally derive the Lorentz transformations does not involve a particularly complicated mathematical formalism, it is beyond the scope of this book. However, we can derive the kinematic dilation of time in a simple way using the well-known Pythagorean theorem and, of course, the principles of Special Relativity mentioned above.

Consider, therefore, a light clock, that is a clock made up of two mirrors placed at a certain distance L and a device that generates a beam of light that reflects in the two mirrors; the unit of time is given by the time taken by the beam to cover the distance between the two mirrors both ways ($2\Delta t'$). Since the distance between the two mirrors is fixed and the speed of light is constant, the aforementioned time is always the same for an observer attached to the clock and $\Delta t' = L/c$. Now, imagine that the clock is moving at constant speed (Fig. 1.2), the stationary observer will see the beam of light following an oblique path, in particular to go from the upper mirror to the lower one and back, it will follow a triangular trajectory. Referring to the right triangle ABC in Fig. 1.2 and applying the Pythagorean theorem we have: $AB^2 = BC^2 + AC^2$, but BC is equal to $L = c\Delta t'$ that is the product of the speed of light for the time interval $\Delta t'$ necessary for the light to go from B to C, and measured by the observer in sync with the clock, $AB = c\Delta t$ that is equal to the speed of light for the time interval Δt necessary for light to go from A to B, as measured by the stationary observer, and finally $AC = v\Delta t$ that is equal to the speed at which the clock moves for the time interval Δt, that is, the distance covered by the clock during the light's journey from A to B. We can then write $c^2\Delta t^2 = c^2\Delta t'^2 + v^2\Delta t^2$, from which it is easily derived that:

$$\Delta t = \Delta t' / \left(1 - v^2 / c^2\right)^{1/2} \text{ or also } \Delta t / \Delta t' = 1 / \left(1 - v^2 / c^2\right)^{1/2}$$

Remember that the superscript ½ indicates the square root. This relationship is called kinematic time dilation, to distinguish it from the gravitational time dilation we will discuss later.

Note that based on the second principle of Relativity we have considered the constant speed both in the reference system in sync with the clock and in the stationary one. The relationship just derived tells us that the time measured by the two observers is different, in particular, the one measured by the

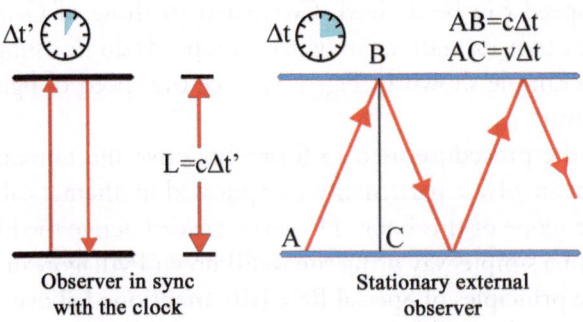

Fig. 1.2 Diagram of Einstein's light clock. Time is marked by the back and forth path of the light beam. For the observer in sync with the clock, time passes more slowly (left). For a stationary external observer, the light travels a longer path and therefore time passes more quickly (right)

observer in sync with the clock ($\Delta t'$), and therefore in motion, is smaller than the one measured by the stationary observer (Δt), since the square root that appears in the denominator of the above equation is always less than 1 as $v/c < 1$.

There is, therefore, a dilation of time, that is, for the observer in motion, time passes more slowly. The time measured in a reference system in sync with the clock is called *proper time* and is always the smallest time that is measured.

We note that if the speed at which the clock moves is negligible compared to that of light, the ratio v/c is practically zero and the time intervals coincide. In everyday life we are always in this last hypothesis, in fact if we wanted to observe a difference of just one second in an hour, using the previously derived relationship of time dilation, we would have to travel for an hour at the incredible speed of about 7000 km/s, consider that the fastest object made by man is the solar Parker probe that in space has reached a speed equal to 95 km/s.

Nevertheless, the imperceptible effects of time dilation have been widely verified even at the macroscopic level. One of the most important experiments was carried out in 1971 by Joseph Hafele and Richard Keating, who, using Cesium atomic clocks placed on the ground and on a plane circumnavigating the globe to the east, measured, after an effective flight time of 41 h, a delay of the atomic clock placed on the plane of about 60 billionths of a second (60 ns) compared to the one on the ground, in excellent agreement with the theoretical predictions (40 ns). This value, significantly higher than the accuracy of the atomic clocks equal to a few ns, also took into account the dilation of gravitational times without which the value would have been about 190 ns.

This experiment leads us to deduce that even the concept of clock synchronization loses its meaning. If you have at your disposal ultra-precise clocks, such as the latest generation atomic ones with an error less than 100 milliseconds in a time of about 14 billion years (equal to the age of the Universe), it seems natural to be able to synchronize them and affirm that they will always report the same time regardless of their position or from their motion. Well, the experiment of Hafele and Keating, reported above, tells us that this is not the case, in fact the passage of time depends on the reference system in which you find yourself: two or more clocks can synchronize only if they are in a reference system in which they are still, i.e. there is no relative motion between them.

However, net of sophisticated experiments that use extremely precise atomic clocks, in the macroscopic world the effects of Special Relativity are not observable, but in the microscopic one, where elementary particles can travel at speeds close to those of light especially in large particle accelerators, relativistic effects are very evident and must be taken into account. In this regard, let's try to consider the case of speeds very close to that of light or at limit exactly equal, we see that the value of the square root that appears in the denominator in the formula for time dilation becomes smaller and smaller, and therefore the ratio $(\Delta t/\Delta t')$ between the time variations for the stationary observer and the moving one is increasingly larger. This means that a few moments for one can also correspond to years or centuries for the other, and if we consider the limit case in which one moves at the speed of light, time stops completely. In other words, if we imagined, as Einstein did, to travel on a beam of light, time would stop: a moment would be equivalent to the whole life of the Universe. Of course, no object with mass, even very small, can travel at the speed of light. Furthermore, also the limit cases in which infinities appear, like the one in which we put $v = c$ in the formula for time dilation reported above, are always very delicate, difficult to interpret and, in some cases, are to be removed in an artificial way as happens in the case of quantum electrodynamics which we will talk about in the next chapter.

Time dilation reassures us about the possibility in the future to carry out interstellar journeys even of a few hundred light years in reasonable times compared to the life of man. If we consider, for example, the planet similar to Earth (TOI 700 d) recently discovered by the TESS telescope and about 100 light years away, a reconnaissance trip to such a planet would last at least 200 years for the inhabitants of Earth, but traveling at relativistic speeds could last even a few years for the astronauts with the not trivial contraindication of likely finding the Earth after two centuries completely changed and not necessarily for the better!

This last consideration introduces us to one of the most famous paradoxes of the theory of Special Relativity, namely the "twin paradox". The first principle of Special Relativity tells us that there is no absolute reference system but all systems of inertial references, that is, those that move in a straight line and at constant speed, are equivalent. It is, therefore, evident that in the mental experiment of light clocks (Fig. 1.2), an observer attached to the clock, that is, who is at rest in the system of reference of the clock, sees the other observer moving and moving away from him at a speed v; therefore, if we put ourselves in the moving reference system we can say that we are still, and the observer who was previously considered still is now in motion. So, we can reverse the argument and say that time is dilated for the observer we previously considered still. This is indisputably true and represents the conceptual basis of Relativity. However, from these considerations arises the apparent paradox of the two twins: one of the two twins leaves for an interstellar journey at relativistic speed, and the other twin stays on Earth, at the end of his journey the astronaut twin finds, due to the dilation of time, the twin brother aged. The paradox lies, in the fact, that for the considerations just made, we could reverse the argument and think that from the point of view of the spaceship, the Earth is not still it is moving at the same speed as the spaceship but in the opposite direction, and it is the spaceship that is still. Therefore, the astronaut twin should find himself aged compared to the brother who stayed on Earth. The paradox is resolved by considering that, when the spaceship returns back to Earth it must decelerate and reverse course and cannot be considered an inertial reference system anymore, so the first principle of Relativity does not apply and the twin who will age faster is the one who stayed on Earth.

From a kinematic point of view, in addition to time dilation another astonishing consequence of Special Relativity and, in particular, of Lorentz transformations is the contraction of space.

Suppose we are on board a spaceship traveling at a certain speed v towards the star Sirius which is 8.6 light years away from Earth (Fig. 1.3) and we measure this distance both from the spaceship and from Earth. For an observer on Earth, both Sirius and the Earth are still while the spaceship is moving at a speed equal to v, therefore if Δt is the time taken by the spaceship to reach the star, the distance will be $D = v\Delta T$. From the point of view of the astronaut, Sirius goes towards the spaceship with a speed equal to v and therefore $D' = v\Delta t'$; if we make the ratio between D' and D we find $D'/D = \Delta t'/\Delta t$. Considering the relation of time dilation $\Delta t'/\Delta t = (1 - v^2/c^2)^{1/2}$, we obtain by simple substitution that $D'/D = (1 - v^2/c^2)^{1/2}$, that is $D' = D(1 - v^2/c^2)^{1/2}$. Since the expression under the square root is always less than 1, objects appear contracted along the direction of motion and the contraction depends on the

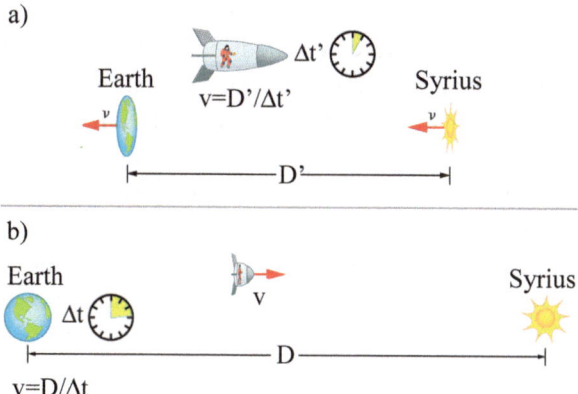

Fig. 1.3 Representation of space contraction. From the astronaut's point of view, Sirius and Earth approach and move away respectively at speed v and the Sirius-Earth distance contracts (a), while for an observer attached to the Earth it is the spaceship that contracts (b). Adapted from Paul Peter Urone, Roger Hinrichs, OpenStax, College Physics, 2012 Texas (licensed under CC BY 4.0)

speed. In other words, a moving object contracts compared to its stationary length.

Even in this case, we hardly see in everyday life cars, trains or planes that contract because the speeds are much smaller than those of light and the effects of contraction are not appreciable. If we consider, for example, a cruising plane that travels at 900 km/h, the contraction measured by an observer on Earth would be just 53×10^{-12} m or a length of the order of the dimensions of a hydrogen atom!

The direct experimental confirmation of space contraction is evidently very difficult to achieve as the variations in lengths to be measured are extremely small, at least at the speeds we are capable of reaching with macroscopic objects. However, there are indirect confirmations like, for example, the high-energy collision of heavy ions in particle accelerators. In this case, some experimental observations can only be explained by considering an increase in electrical density due to length contraction: in fact, if the lengths contract, the volume decreases and being the charge always the same, there is an increase in charge density defined as the ratio between the total charge and the volume.

As for time, the length measurement that is made in one's own reference system is called *proper length* and is always the greatest length that is measured.

To conclude the aspects of relativistic kinematics, it is necessary to mention a concept that takes on another meaning or rather it would be more correct to say that it loses its meaning in the context of the theory of Relativity: simultaneity. Two events are defined as simultaneous if they happen at the same instant. If we imagine being in the center of a train cabin that travels at a constant speed and simultaneously, we start two light beams, these for an observer attached to the cabin, will arrive simultaneously at the beginning and at the end of the cabin. It can therefore be stated that in a reference system whether it is at rest or in uniform rectilinear motion, we can define the simultaneity between two events. However, for an observer stationary at the station the events are not simultaneous, as being the speed of light equal in all reference systems, from the station you see the two light beams traveling in opposite directions, but since the train is moving (let's assume to the right), the beam of light traveling to the left will meet the end of the cabin a moment before the beam traveling to the right. So, for the observer stationary at the station the events are not simultaneous. Simultaneity like time and space takes on a relative character, i.e. it depends on the reference system. Obviously, the effect depends on the speed with which the reference systems move between them and, in this particular case, also from the spatial distance in which the two events occur, in the case of the train example, this distance is the length of the cabin.

1.2   Energy from Mass

Probably the most disruptive aspects of the theory of Special Relativity, especially from a point of view of applications, are those related to dynamics or the study of causes that determine the motion of bodies.

In 1687, Isaac Newton, considered alongside Einstein as one of the great geniuses of physics, published a treatise in three books titled *Philosophiae Naturalis Principia Mathematica* (The Mathematical Principles of Natural Philosophy), which, according to many, is one of the most important scientific works. In the first book, *De Motu Corporum* (On the Motion of Bodies), Newton introduces the three principles of dynamics on which all classical mechanics is based. The first principle, also known as the principle of inertia, essentially states that the natural state of a body is rest or uniform rectilinear motion (at constant speed), completely in contrast with Aristotle's concept of motion, according to which the motion of a body is always determined by a motor in contact with the body. The second principle, typically expressed with the famous formula $F = ma$, states that if a constant force is applied to a body, the latter moves with an acceleration that is directly proportional to the force and inversely proportional to its inertia quantified by an intrinsic characteristic of the body, namely its mass. Finally, the third principle, known as the principle of action and reaction, states that in a system in which no external forces act, the force that one body exerts on a second body is equal and opposite to the force that the second body exerts on the first.

For each of these principles, it would be worth elaborating to appreciate their importance, however let's limit ourselves to some observations on the second principle: in principle if a force is applied to a body and we can neglect friction and air resistance (one can imagine a body in interstellar space), the body begins to move with a constant acceleration and, consequently, since acceleration is nothing more than the increase in speed per unit of time, it follows that the speed increases linearly with time. Therefore, after a certain time that can be easily calculated, the body will have reached the speed of light and in the next instant it will have exceeded it, which is not possible according to the second principle of the theory of Special Relativity that sets the speed of light as the maximum speed of a body. It is therefore evident that the formula expressing Newton's second principle needs to be changed with a more general one that takes into account the second principle of Special Relativity. A similar argument must be made for the other fundamental quantities of dynamics: kinetic energy and momentum in whose expressions the speed explicitly appears. In the case of kinetic energy ($E_c = 1/2mv2^2$), which is a measure of the energy of a body related to its motion, it is proportional to the mass m and the square of the speed V, implying that a 40% increase in speed results in a doubling of kinetic energy. Also in this case, it is possible from a classical mechanics point of view to provide a body with such energy to exceed the speed of light. The same argument applies to the other fundamental quantity of dynamics, momentum equal to $P = mv$. In light of this consideration, it became clear to Einstein that it was necessary to reformulate Newtonian dynamics by finding more general expressions for force, energy and momentum that would reduce to the classical ones for speeds much smaller than that of light.

A few months after the first article on the theory of Relativity, Einstein published another article titled "Does the inertia of a body depend on its energy content?", which we can consider a sort of addendum to the theory of Special Relativity in which, using the principles of the theory and the well-known laws of classical mechanics, he arrives at the most famous formula in physics, which in its simplified form (reference system in which the body is stationary) takes the expression $E = mc^2$, where m is the mass of the body and c is the speed of light. Here is the other revolutionary concept of the theory of Relativity, both from a theoretical and practical point of view: mass is no longer an independent and absolute quantity but is a form of energy; therefore, it is possible to convert mass into energy and vice versa. Therefore, the principles of conservation of mass and energy must be merged into a single, more general principle of conservation, that of mass-energy. The famous principle of the great French scientist Antoine Lavoisier, according to which in a chemical reaction the sum of the masses of the reactants is equal to the sum of the masses of the products, is true only in the approximation in which we can neglect the mass that transforms into energy.

But if mass is a form of energy and the famous formula written above holds, the transformation into energy, even of a very small mass, can produce a huge amount of energy: the integral transformation of one gram of matter would produce an energy equivalent to about 10^{14} Joule, equivalent to the electrical energy consumed by a family in about 1000 years!

Einstein's famous formula has had a scientific and technological impact of enormous scope. The huge amount of energy released in the processes of nuclear fission and fusion is based precisely on the transformation of mass into energy.

In nuclear fission, a radioactive atom (typically uranium or plutonium) following a collision with a neutron splits into two lighter atoms whose total mass is less than that of the starting atom; the difference in mass, also known as the *mass defect*, transforms into energy acquired by the reaction products (atoms and neutrons). If the amount of fissile material exceeds a certain critical mass, the spare neutrons can induce other fissions of uranium or plutonium atoms, triggering a chain reaction. This is the phenomenon on which nuclear power plants are based, where the chain reaction is controlled by moderation rods (typically made of graphite) that, by capturing neutrons, manage to prevent overheating of the reactor and above all that the chain reaction becomes explosive as happens in nuclear bombs. In addition to the risk of explosions, the problem with nuclear power stations is the radioactive waste, which can take from 20 to 300 years to deplete.

In the reverse process, nuclear fusion, two light atoms, deuterium and tritium (hydrogen atoms with one and two neutrons, respectively), fuse and form a helium atom and a neutron. Once again, the sum of the masses of the reaction products is less than that of the sum of the masses of the starting atoms and the difference in mass transforms into energy. Unlike fission, there are no radioactive waste nor the danger of an uncontrolled chain reaction; however, the technology to build a fusion power plant is much more complex. For a fusion process to occur, temperatures of about 150 million degrees and enormous pressures are needed, as atomic nuclei being positively charged repel each other with a force (Coulomb force) that increases more and more as the nuclei get closer; only when the distance between them is extremely small (on the order of 10^{-15} m) the repulsion force is compensated by a stronger attractive force, the strong nuclear force, which allows protons and neutrons to stay close in atomic nuclei without repelling each other. The research currently underway to achieve thermonuclear fusion is essentially of two types: inertial confinement, in which powerful laser beams heat and compress the mixture of tritium and deuterium with such pressure as to induce the fusion process, or magnetic confinement, in which very powerful magnetic fields generated by superconducting magnets confine the plasma of tritium and deuterium in a toroidal geometry, and the heating of the plasma to the temperature useful for fusion occurs through a current induced by external coils. After 50 years of enormous efforts directed at this strategic field of research, on December 13, 2022, the United States Department of Energy has announced an extraordinary result that bodes well for the future of the entire planet. For the first time, the energy produced by an inertial confinement-based nuclear fusion reaction was greater than the energy needed to trigger the reaction itself. Specifically, about 2 million Joules of energy used by lasers to produce nuclear fusion resulted in about 3 million Joules, a gain of about 50%.

Encouraging results in the field of nuclear fusion have also been achieved with magnetic confinement technique. In fact, in 2023, the Japanese *tokamak*

(JT-60SA) was able for the first time to achieve magnetically confined plasma inside its toroidal chamber. This result is very important in view of the realization of the mega reactor *ITER* (acronym for International Thermonuclear Experimental Reactor), under construction in France, and due to an international collaboration in which Italy plays a significant role.

The Sun and other stars can be considered large thermonuclear power source that transform mass into energy allowing life on Earth and possibly on some other planet in the Universe. Specifically, in the Sun, fusion reactions of hydrogen atoms into helium primarily occur. If mass transforms into energy, do the Sun and other stars lose mass and, therefore, are destined to burn out? Actually, the death of a star occurs long before it is completely consumed by the transformation of mass into energy. In the particular case of the Sun, the average radiated power is 4×10^{26} W, which corresponds to a decrease in mass of 4.4×10^9 kg every second; given that the mass of the Sun is about 10^{30} kg, it would take over 1000 billion years to transform 10% of the Sun's mass into energy. Our star, according to the latest estimates from the European Space Agency (ESA), is 4.6 billion years old and is halfway through its life, it will therefore die long before 10% of its mass is transformed into energy.

Once the hydrogen inside the Sun's core is exhausted, in about 5 billion years, nuclear fusion reactions will involve heavier atoms that will fuse to produce carbon and oxygen. This phase corresponds to a strong instability that will lead to a great expansion of the Sun becoming a *red giant* with a diameter equal to the entire solar system. Once the fusion processes are finished, there will be a strong downsizing until it becomes a *white dwarf*, very dense and not very bright and with a diameter smaller than that of the Earth. What will be the fate of the Earth? During the Sun's expansion phase into a red giant, it will be reduced to a huge burnt rock devoid of any form of life, but most likely by that time life on Earth will have become extinct for other reasons.

Einstein's famous formula and more generally the theory of Special Relativity, as we will see later, also plays a fundamental role in quantum physics and, in particular, in theories of elementary particles and fundamental interactions.

1.3   General Relativity: The Geometrization of Gravity

Einstein's greatest work, which places him among the greatest geniuses of humanity, is undoubtedly the theory of General Relativity or the new theory of gravitation. The theory of Special Relativity, even without Einstein, would probably have been born anyway, there were too many experimental clues leading in that direction. The theory of General Relativity, on the other hand, is the result of a brilliant vision that had no experimental clue requiring its existence, except for a very small deviation of the precession of Mercury's perihelion that could not be explained within the framework of Newtonian mechanics.

Before proceeding with the most beautiful theory of physics, as considered by many scientists, we must dwell on a concept that will also be present in the rest of the book, namely the concept of *field*.

Rather than giving a formal definition of field, we will provide examples that help to effectively understand this important concept of physics. Suppose we are in a room in winter with heat sources to warm us (electric heaters or radiators) and to measure the temperature at all points in the room. Obviously, we expect that the temperature is not the same everywhere, in particular near the heat source it is higher and, as we move away, it is lower. We can therefore say that every point in the room, including the walls and the floor, is characterized by a precise value of the temperature, that is, there is a temperature field in the room. In this case, we speak of a scalar field since temperature is a magnitude *scalar* that is, it requires only one value (intensity) to be completely defined. Now suppose we are able to measure the gravitational force that the Earth exerts on a sample mass of 1 kg in the Earth's atmosphere. If we were able to make this measurement at infinite points in the atmosphere we would observe a nearly constant value. However, if we exceeded the atmosphere and repeated the measurement a few thousand km from the surface of the Earth, we would observe that the value of the force of gravity decreases as we move away from the Earth. So also in this case, we can affirm that all the space surrounding the Earth is the site of a field of gravitational forces but in this case we talk about a field *vector*, since a force is completely defined if we provide the intensity, direction and sense. In other words, to define a force we need an entity geometric known as *vector*, that is, a segment oriented in the space whose length provides us with the intensity of the physical quantity it represents. The same argument can be repeated with any other force such as electric, magnetic or nuclear. In conclusion, we can affirm that a field is a property of space or, better still, of space-time, that is, each point is characterized by one or more physical quantities that define the various fields. What has been said leads us inevitably to deduce that the space that surrounds us is always the site of various fields (gravitational, electromagnetic, thermal, etc...). As we will see in the next chapter, also in intergalactic spaces in which apparently there is absolute vacuum, there are particles and fields.

Let's now return to the theory of General Relativity trying to understand the motivations that led the German genius to develop a more general theory and how gravity came into play.

Einstein was allergic by nature to accepting any dogma whether it came from men or apparently from nature. The idea that there were privileged reference systems like those studied in Special Relativity disturbed him a lot. He believed that laws of physics should preserve their form in all reference systems including those non-inertial, that is, systems that move with arbitrary motion (constant speed, constant or variable acceleration), thus generalizing the first principle of Special Relativity. As stated in Einstein's article on General Relativity of 1916: "The laws of physics must be of such a nature to be valid with respect to a reference system in arbitrary motion. We arrive in this way at an enlargement of the postulate of Relativity."

Driven therefore by the need to generalize his theory of Relativity also to the case where the reference systems were not inertial, in 1907 Einstein intuited what he himself defined as "the happiest thought of my life" that is, that a body in free fall does not feel its own weight or, in equivalent form, there is no difference between a gravitational field and an accelerated system (principle of equivalence). During the Kyoto conference in 1922, Einstein recalls: "I was sitting in an armchair in the patent office in Bern when suddenly I found myself thinking: If a person falls freely, they do not feel their own weight. I was astounded. This thought, so simple, struck me deeply, and I was propelled towards a theory of gravitation".

To try to understand the meaning of the principle of equivalence, let's imagine being in a cabin of a spaceship without windows, completely sound-proofed and in which we have set up a room with a table, chairs and other objects in a way that is completely identical to a room in an apartment on Earth. If the spaceship begins to move with an acceleration upwards equal to the acceleration of Earth's gravity g, the people, the table, the chairs and all the objects in the room are subjected to an acceleration equal and opposite to that of the spaceship and therefore directed downwards. A similar situation is observed when the car or train accelerates, or even more when the plane is in phase of takeoff, one feels a force that pushes us towards the back of the seat. These phenomena are widely known in classical physics and the forces that determine these phenomena are called *apparent forces or fictitious forces* and are due to the accelerations of reference systems.

Returning to the spaceship, when it has sufficiently moved away from Earth so as to be able to neglect the force of gravitational attraction, in the arranged room there will be only acceleration equal and opposite to that of the spaceship, there is therefore an acceleration directed towards the floor of the room

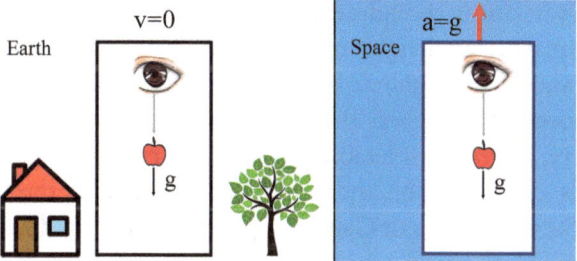

Fig. 1.4 Principle of equivalence. An observer on Earth sees an apple fall with the acceleration of Earth's gravity (9.81 m/s²). An observer in a spaceship that is moving with an acceleration equal to Earth's gravitational acceleration, sees the apple fall exactly as the observer on Earth

equal to g,. for which reason whatever one does (throw a ball, move the table and chairs, jump around the room or drop an apple) one has the impression of being on Earth as everything is attracted towards the floor just as it happens on Earth (Fig. 1.4).

In other words, there is no way to notice if one is on a planet in a room without windows or on a spaceship without windows that travels with a constant acceleration equal to the gravitational acceleration of the planet. An inertial force, due to the acceleration of the reference system, can therefore simulate a gravitational force.

The principle of equivalence also tells us that a body in free fall does not feel its own weight, the meaning of which can be understood with the famous thought experiment of Einstein's elevator. Imagine being in an elevator and that, unfortunately for the person inside, the elevator cable suddenly breaks, during the fall the passenger begins to float in the void like an astronaut, since the fictitious force, due to the acceleration downwards of the elevator, is directed upwards and being equal and opposite to the gravitational attraction force it cancels it out, giving weightlessness to all objects present in the elevator. In this case an inertial force due to the acceleration of the reference system can locally cancel a gravitational force. Einstein understood therefore that the generalization of Relativity would have implied a new concept of gravity.

Let's take a step back to understand what the conception of the gravity before Einstein. In his "Dialogues on the Two Chief World Systems," Galileo had discussed the fall of heavy bodies, highlighting that the acceleration with which bodies fall is the same regardless of their mass, unlike Aristotle's theory of gravity, according to which objects fall at a speed proportional to their mass. It is said that Galileo would have verified his hypothesis with the famous experiments carried out from the Tower of Pisa, but most likely that of the

Tower of Pisa was only a thought experiment. In fact, a feather and a lead ball, if dropped from a certain height, touch the ground at the same time, provided that air resistance is eliminated. In reality, the lead ball touches the ground decidedly before the feather, as the gaseous fluid in which we are immersed, the atmosphere, offers considerable resistance to falling bodies and this resistance depends exclusively on the aerodynamic shape of the bodies. But to easily realize the correctness of Galileo's hypothesis, we can do a simple experiment. Take an A4 sheet of paper and a fairly heavy book with dimensions similar to that of the A4 sheet (29 cm × 21 cm), if we take the sheet with one hand and the book with the other and let them fall to the ground at the same time, we notice that the sheet starts to flutter and touches the ground after the book. This is because air resistance affects the sheet of paper more, which has a much smaller mass than the book. If we now place the sheet of paper on top of the book and let the book fall with the sheet resting on top, we observe that the two objects reach the ground at the same time. This is due to the fact that the book in front of the sheet greatly reduces the air resistance on the sheet.

Numerous accurate experimental tests for the verification of the fall of heavy bodies have been carried out, confirming Galileo's hypothesis with extreme precision. One of the most spectacular experiments was carried out during the Apollo 15 mission on the moon in 1971. Astronaut David Scott dropped a hammer weighing over 3 kilos and a feather weighing about 3 grams from his hands, observing, as also witnessed by the historic footage taken by the other astronaut James Irwin, that the two bodies fell at the same speed and touched the ground simultaneously. There is no atmosphere on the moon and therefore no air resistance; therefore, it is the ideal place to directly verify the law of the fall of heavy bodies.

Galileo's results were fundamental and allowed a huge step forward in understanding this force with which we deal every day, but the formalization also in mathematical terms of the theory of gravitation was developed about 100 years later by Newton who intuited that the same force that makes the famous apple fall from the tree also makes the moon orbit the Earth and all the planets orbit the Sun. In the third book of the *Philosophiae Naturalis Principia Mathematica*, Newton states the famous law of universal gravitation according to which two bodies attract each other with a force that is proportional to the product of their masses and inversely proportional to the square of their distance. Since the proportionality constant is very low, we typically do not observe that the table in our living room attracts the chairs as the friction forces at play are decidedly more intense than the weak gravitational interaction. The effects of attraction between bodies are only seen when the masses involved are very large, like that of the Earth or the Sun. The law of universal gravitation and the second law of dynamics also explain Galileo's law of the fall of heavy bodies provided that it is assumed that the gravitational mass, that is, the one sensitive to the force of gravity, is equal to the inertial one, that is, the mass that opposes motion when we exert a force on it. This assumption is known as the principle of equivalence between inertial and gravitational mass and is experimentally verified with very high precision.

Despite the extraordinary predictive capacity of the universal law of gravitation, especially in the field of celestial mechanics, Einstein had a hard time digesting the instantaneous action predicted by Newton's famous law. The second principle of Relativity predicted a speed limit (that of light), while according to Newton's theory of gravitation, the attraction between two bodies is instantaneous, even if they are very far apart from each other. In short, for Einstein it was absurd to imagine that, if the Sun disappeared, the Earth would immediately leave its orbit, but it was plausible to imagine that it would do so after no less than 9 min, that is the time it takes for light to reach the Earth from the Sun.

Let's return to Einstein's thought experiments and consider an observer in an elevator moving upwards with a high acceleration rate; if from the outside a beam of light is sent horizontally into the elevator through a suitable hole, the observer inside the elevator will see the beam of light curve downwards since, while the beam travels horizontally through the elevator covering a distance proportional to time, the elevator rises, covering a vertical distance that is proportional to the square of the elapsed time (Fig. 1.5). Since a beam of light appears curved when watched by an observer in the accelerated system, by the principle of equivalence, Einstein deduced that the gravitational fields

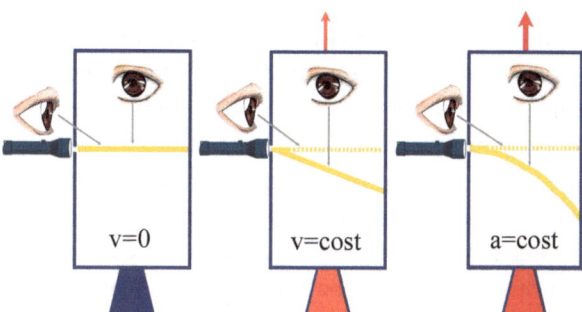

Fig. 1.5 A beam of light is introduced into the rocket by an external observer, who sees the light proceeding in a straight and horizontal manner (dashed line). The trajectory of the light represented by solid line is that seen by the observer inside the rocket. For large accelerations, the light follows a curved path and by the principle of equivalence behaves in the same way in the presence of large masses

generated by planets and stars, like accelerated systems, must curve light beams in their vicinity, that is, gravity curves light beams.

But if light near large masses bends, then the space crossed by the light is curved, therefore gravity produces a curvature of space-time.

We therefore move from a vision of a force at a distance (Newton's law of universal gravitation) to a new revolutionary vision that predicts the geometrization of gravity: masses curve space-time allowing other masses or light to slide into these depressions of space-time (Fig. 1.6). We can therefore imagine space-time as a stretched sheet, if we put a huge lead ball on it, the sheet will deform assuming a funnel shape and another body near the lead ball will follow the depression of the sheet and will move towards the huge lead ball. Therefore, objects move along the shortest path in a curved space-time. The shortest line that connects two points (*geodesic*) is no longer a straight-line segment as happens in flat Euclidean space but a curved line that changes depending on the curvature of space-time. The greater the mass of the body, the more pronounced is the deformation of the space-time fabric (Fig. 1.6). As the astrophysicist American physicist John Archibald Wheeler said: "matter tells space how to curve, space tells matter how to move". However, the The general theory of Relativity tells us that the curvature of space-time is proportional to the density of matter and the ratio between the universal constant of gravitation ($G = 6.7 \times 10^{-11}$ N m^2/kg^2) and the square of the speed of light ($c = 3 \times 10^8$ m/s). Therefore, being the aforementioned ratio extremely small (about 7×10^{-28} m/kg), the effects of space curvature are appreciable only in the case of very large masses (stars, galaxies).

Essentially, gravity is nothing more than a property of space-time and the force of gravitational attraction is due to its deformation produced by massive bodies.

In more formal terms, the *metric,* which determines the distance between two points in space-time is nothing more than the gravitational potential. In a flat Euclidean space, that is the space with which we essentially deal every day, the metric is Pythagorean and the distance between two points is simply the straight line that joins the two points and is calculated by applying the Pythagorean theorem. It is easy to imagine that in the case of a curved space, the distance between two points is not so simple to calculate.

The intuition of the geometric nature of gravity that occurred in 1912 was a further and perhaps more important step towards the theory of General Relativity. However, these incredible intuitions had to be concretized in a physical theory that had to be written in the natural language of physics or that of mathematics and confirmed by experiments. Dealing with curved spaces, the mathematical apparatus became much more complicated. It

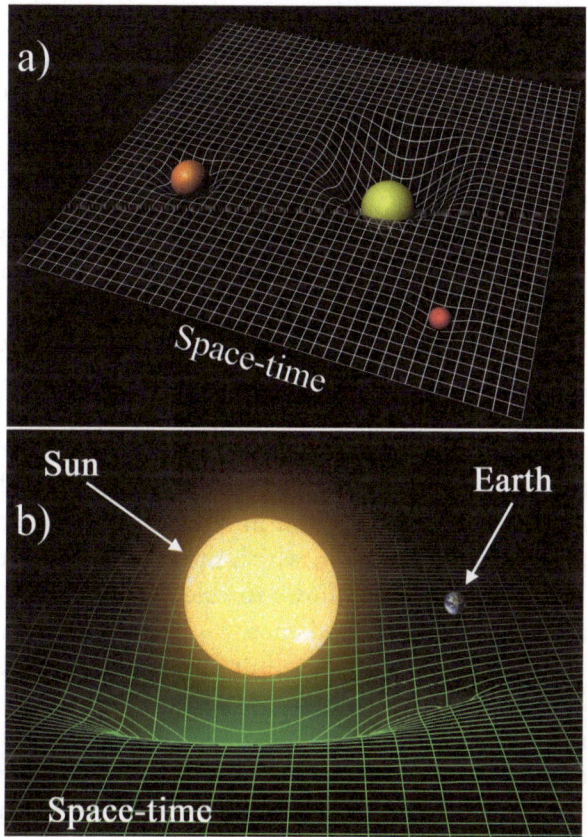

Fig. 1.6 Artistic illustration of the deformation of space-time in the presence of massive bodies. (**a**) A greater mass corresponds to a greater deformation of the space-time fabric. Credit: ESA - C. Carreau. (**b**) Illustration of the Earth's orbit around the Sun. Our planet follows the deformation of space-time produced by the Sun. Credit: Caltech/MIT/LIGO Lab/(T.Pyle)

required the non-Euclidean geometry developed by Carl Friedrich Gauss and Bernhard Riemann and above all the differential calculus on curved surfaces. Einstein did not have that knowledge of advanced mathematics and realized that the path was difficult, long and tiring, and then asked for help from his friend Marcel Grossman (mathematician and colleague of Einstein at the University of Zurich); his desperation is well described by the appeal to his dear friend in 1912 "Grossman help me or I'll go crazy". Grossman directed him to studies on the "absolute differential calculus" conducted by great Italian mathematicians and in particular by Luigi Bianchi, Gregorio Ricci Curbastro and Tullio Levi Civita. Having mastered the mathematical formalism, Einstein immediately manages to write the equation of motion of a body (geodesic equation) in a given gravitational field or equivalently in a given space-time, whose solution is no longer a straight trajectory, but a curve and, in the case of weak fields, the equation is reduced to Newton's equation. But the most arduous aspect of the theory was undoubtedly understanding how a certain distribution of mass-energy determined the characteristics of space-time and therefore of gravity. He then begins the final step aimed at the elaboration of a general equation that contained the formidable intuitions of the previous steps and, at the same time, included as a limiting case the theory of Newton.

After a long period of hard work, dedication and perseverance, the German genius, with four lectures held at the Prussian Academy of Sciences in Berlin, November 1915, exposed the theory of General Relativity, which essentially replaced and incorporated Newton's theory of universal gravitation. The results were then published on December 2, 1915 and in a more extended version in March 1916, in an article titled "The basis of General Relativity" in which the fundamental equation takes the name of *field equation* and allows to calculate the geometry of space once the distribution of masses and energy is known. Note that we are not only talking about mass distribution but also of energy since, as seen in the previous paragraph, mass and energy are equivalent.

Even though the description of the mathematical and formal details goes beyond the scope of this book, given that Einstein's field equation is one of the most important and beautiful equations in physics, it is worth mentioning, without any intention of delving into technical formalisms suitable for advanced university courses. To not burden the reader, we report it in a rather singular way, that is, painted by an unknown physics enthusiast on an abandoned locomotive in the Atacama desert in Chile (Fig. 1.7).

The symbols with Greek subscripts are mathematical objects called *second order tensors*, the Greek indices can take 4 values (0,1,2,3), that is, the four dimensions of space-time. The tensors $R_{\mu\nu}$, $g_{\mu\nu}$ that appear on the first member of the equation are the Ricci tensor and the metric tensor respectively and take into account the deformation of space-time while the one on the second member ($T_{\mu\nu}$) is the energy-mass tensor that takes into account the distribution of mass and energy that deforms space-time. On the first member we also find the *scalar curvature R* also linked to the curvature of space-time. Finally, G is the universal gravitational constant and c is the speed of light. Since the indices can vary from 0 to 3, there are sixteen possibilities to combine them which are reduced to ten as the tensors are symmetric ($R_{10} = R_1$, $R_{12} = R_1,\ldots g_0 = g_1, g_2 = g_{21},\ldots T_{10} = T_{01}, T_{12} = T_{21},\ldots$).

Fig. 1.7 Einstein's field equation painted on an abandoned train locomotive in the Atacama desert in Chile

So, in reality these are ten differential equations written in a compact way whose exact solution, as one can easily guess, is not at all easy. Of course, Einstein, as the first thing, shows that in the case of weak fields, that is, in cases where space is almost flat, the field equation reduces to the classic one of Newton.

The path that leads to the field equation and, therefore, to the conception of the final theory lasts eight years, during which various attempts were made and, in the final part of the difficult journey, Einstein feared the competition with one of the greatest mathematicians of the time, David Hilbert, who, after inviting Einstein to give a series of lectures on General Relativity, in Gottinger between 1913 and 1914, had started to deal with the problem and being a great mathematician was a formidable competitor for Einstein.

That was one of the most stressful periods from the point of view psychological for Einstein. In fact, Hilbert published in December 1915, almost simultaneously with Einstein (for some even earlier), an article with the correct field equations but acknowledged in his article the paternity of the theory to Einstein: "The differential equations of gravity obtained seem to me in agreement with the magnificent theory of General Relativity enunciated by Einstein in his latest article". During the discussions with Hilbert, Einstein expressed his doubts also related to mathematical difficulties and Hilbert argued that "Physics is too difficult for physicists", wanting to underline the difficulty in using advanced mathematics.

1.4   Predictions, Experimental Verifications and Cosmological Implications

The theory of General Relativity presented in the first article of 1916 was able to account for the small deviation of the precession of Mercury's perihelion and above all predicted quantitatively how much the Sun deflected a light ray. For Einstein, the exact prediction of the precession of Mercury's perihelion was very important as it demonstrated that the theory worked and, in fact, he was very happy with this first result confirmed by experimental data. By precession of the perihelion of Mercury's orbit, we mean the rotation (precession) of the point closest to the Sun (perihelion) of the planet Mercury's orbit. Among all the planets of the solar system, Mercury is the one that presents the most pronounced precession of the perihelion, being the closest to the Sun. In other words, the point of the orbit closest to the Sun does not remain still but rotates in an extremely slow manner, in particular it moves by 5600″ (arc seconds) every century, that is about 1.5 degrees per century. Remember that one degree is worth 60 arc minutes and one minute 60 arc seconds.

In 1859, the famous French mathematician and astronomer Urbain Jean Joseph Le Verrier, based on the Newtonian interaction between Mercury and the other planets, predicted a shift of 5557″ per century with a deviation of 43″. Le Verrier had a deep knowledge of celestial mechanics and was famous especially for his contribution to the discovery of Neptune, using only mathematical calculations and previous astronomical observations. The differences between the observed orbit of Uranus and that predicted by Kepler's and Newton's laws could only be explained if the existence of another planet (Neptune) was hypothesized, discovered by Johann Galle in 1846. In the same way, the great astronomer believed that that deviation of 43″ could be explained by hypothesizing the existence of another planet, which was named Vulcan in anticipation of its discovery. In reality, the planet Vulcan was never discovered and in 1915 Einstein announced that the theory of General Relativity predicted a precession of the orbits of the planets that had to be added to that due to the interaction between them and that the magnitude of this precession for Mercury corresponded exactly to the observed deviation and therefore it was not necessary to hypothesize the existence of another planet.

But the most important prediction of the theory of General Relativity was the deviation of light rays by the Sun (Fig. 1.8), according to calculations made by Einstein this deviation amounted to 1.75″ arc seconds (less than one thousandth of a degree). An experiment that confirmed this data would have removed any doubt about the validity of the theory. Obviously, to be able to

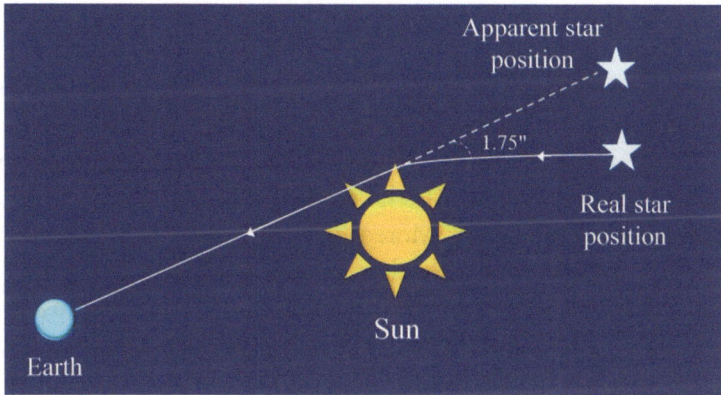

Fig. 1.8 Deviation of light rays coming from stars by the Sun. The angle of deviation is 1.75 arc seconds, i.e., less than one thousandth of a degree. In order to make the effect of the curvature of light rays visible, a deviation angle of about 30° is shown in the figure

observe a ray of light deflected by the Sun, a solar eclipse was necessary otherwise the Sun's rays overwhelmingly covered the weak light rays coming from other stars and weakly deflected by the Sun. The great success came in May 1919, when the English astronomer Sir Arthur Eddington, during the solar eclipse managed to measure from two different points of the globe the deviation by the Sun of the light emitted by the stars. In particular, Eddington led two expeditions: one to the island of Principe, near the Gulf of Guinea in equatorial Africa, and the other to Sobral, a Brazilian municipality located in the northeastern part of the country.

That day, the Sun was shining in the constellation of Taurus, among the stars of the Pleiades, a large star cluster that provided the ideal background for the eclipse and about 150 light years away from Earth. The Eddington expedition measured an actual deviation of the light rays coming from the Pleiades, a deviation of 1.61″ from Principe and 1.98″ from Sobral was observed, in excellent agreement with the predictions of General Relativity (1.75″). The measurement was repeated in 1922 and in 1973 with results consistent with those of Eddington.

After this experiment, Einstein became a famous figure, a pop icon ante litteram, the most important newspapers in the world talked about the German scientist and his scientific revolution. Time magazine in 1999 dedicated the cover to him and considered him "person of the century".

If large masses produce a deviation of light rays, is it possible to observe on a cosmological scale similar effects to what happens when a light ray passes through an optical lens (magnification or duplication of an image)? In other words, are there conditions to observe a gravitational lensing effect, better known as *gravitational lensing*?

On this aspect, Einstein was stimulated by his friend, Rudi W. Mandl, an engineer and physics enthusiast, who in 1936 asked him to evaluate the possibility of a noticeable deflection of light such as to observe gravitational lensing. Einstein was very skeptical as he tended to be quite conservative about the consequences of his theory. Moreover, as early as 1912 he had considered this possibility and had concluded that the effects were negligible; however, he redid the calculations and published the results in a brief note in the journal *Science*, in which he essentially excluded the possibility of observing deflections of light such as to produce an image with the same effects as an optical lens. But in 1937, the Swiss astronomer Fritz Zwicky in two articles argued that the gravitational lens effect can be seen using galaxies with masses much larger than the individual stars to which Einstein referred in his article. Therefore, if a celestial body is located near a huge distribution of matter, like a black hole or a galaxy, the light emitted by it is deflected as happens with an optical lens (Fig. 1.9). The deflection produced is proportional to the mass of the body that acts as a lens and is inversely proportional to the minimum distance at which the light ray passes from the lens itself.

In fact, Zwicky was not wrong, but to see the first images generated by a gravitational lens, one had to wait more than forty years. In 1979, D. Walsh, R.F. Carswell and R. J. Weyman, using a telescope installed in an observatory

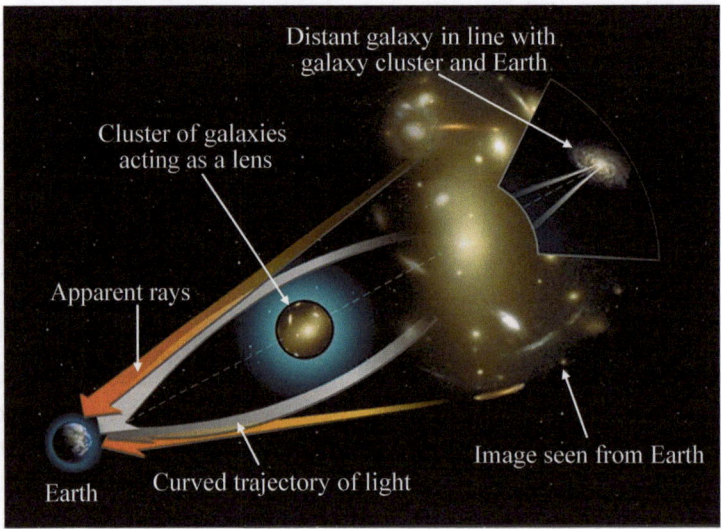

Fig. 1.9 Representation of a gravitational lens. A cluster of galaxies can produce a deflection of light rays such as to produce a magnification or duplication of images

in the Arizona desert, visualized an image of two quasars (very bright galaxy nucleus) only 6 arc seconds apart. It was the same celestial object whose image had been doubled due to the gravitational lens effect caused by the galaxy interposed between the Earth and the quasar.

The *gravitational lensing* has become a valuable tool in the field of astrophysics and cosmology.

It is used, for example, to determine the characteristics, such as mass and distance, of celestial objects that act as lenses and/or of light sources, allowing to identify very distant objects, typically galaxies, that would otherwise not be intercepted in any other way. In 2021, the Hubble Space Telescope identified, thanks to *gravitational lensing*, a distant galaxy 10 billion light years away from Earth. Another important application of *gravitational lensing* is the study of dark matter/energy which represents over 90% of the matter of our Universe and which cannot be seen as it does not emit and does not reflect electromagnetic radiation, but its gravitational effects are unequivocally observed. Dark matter deflects light coming from distant galaxies, causing a small deformation of the images of the galaxies, from which the distribution of dark matter in that particular area of the Universe under observation can be inferred. In this regard, in a study published in Science in 2020 and mainly conducted by a team of astrophysicists from the National Institute of Astrophysics, data from the Hubble Space Telescope and the Very Large Telescope of the European Southern Observatory in Chile were collected. Using the *gravitational lensing* it was shown that dark matter seems to have a gravitational effect at least ten times more intense than expected.

In the first paragraph we saw that one of the most important consequences of special relativity is time dilation. It is therefore natural to ask: what happens to time in the context of the General Relativity theory? Does gravity affect time? To understand this rigorously we would need to make mathematical considerations on the field equation that go beyond the scope of this book. However, if we consider Fig. 1.6, it is immediately clear that the space-time deformation depends on the point where it is evaluated: it is greater near the mass and as we move away it becomes less and less pronounced until it almost completely disappears at great distances from the mass that produced the deformation. This implies that temporal events, which can be considered particular cases of geodesics between two points of space-time where the spatial coordinate is the same, depend on the point where they are considered and this means that the times measured by observers placed at different points in space-time are different. There is therefore a gravitational time dilation and in particular time runs slower in points where the deformation is greater, i.e. near large masses. In other words, time runs slower where gravity is more intense, we call it gravitational time dilation to distinguish it from the kinematic time dilatation seen in the first paragraph. Therefore, time on Earth runs slower than that on a spaceship in orbit. Even in this case, the effects are imperceptible and depend on the mass of the planet where we are. For example, an astronaut in orbit at 400 km altitude for 6 months and at a speed of 8 km/s, ages by a billionth of a second more compared to the inhabitants on Earth!

Numerous experiments have shown gravitational time dilation. In addition to the already mentioned experiment by Hafele and Keating, in 1976 Luigi Briatore and Sigfrido Leschiutta measured with cesium atomic clocks a difference of 30 billionths of a second per day between the clock placed on the top of the Plateau Rosa, at 3500 meters above sea level, and one placed in Turin at 250 meters above sea level. This implies that, living an entire life (80 years) on the Rosa Plateau, one would age about 1 ten-thousandth of a second more than in Turin! In an extraordinary experiment in 2010 published in Science, gravitational slowing was even measured on a difference in height of just 50 centimeters. The current ytterbium atomic clocks with a precision of one part in 10^{18} (equivalent to an error of about one-tenth of a second over 14 billion years), would be able to appreciate the gravitational slowdown even of a few centimeters of difference in height or the kinetic one of a few m/s.

Even though these effects are really small enough to make us think that they will never have any effect on everyday life, in reality it is not quite so. With the advent of the Global Position System (GPS), it is possible to reach a destination with a precision of a few meters and it is now a navigation tool available on smartphones and widely used in everyday life. The GPS is a positioning system based on orbiting satellites, capable of providing the exact position and time to any device equipped with a suitable receiver. The satellites orbit around the Earth at a height of about 20,000 km and at a speed of about 4 km/s; therefore, they accumulate a delay due to cinematic time dilation of about 7 millionths of a second per day and an advance of about 46 millionths of a second per day due to gravitational time dilation, that is a total advance of 39 millionths of a second per day. Considering that the signals emitted by the satellites for localization travel at the speed of light, the localization error is simply given by the speed of light for the time advance or about 12 km in one day! It is therefore necessary to make the necessary corrections otherwise on a journey of about 6 h we could find ourselves in a place that is 3 km from the set destination. In practice, the correction is made by slowing down the onboard clocks, the nominal frequency of 10.23 MHz is changed to 10.22999999545 MHz.

Beyond this useful and important application, it is reasonable to imagine, from what has been said so far, that the most important and significant implications of the theory of General Relativity are in the astrophysical and cosmological field.

At the beginning of 1916 the German mathematician and astrophysicist Karl Schwarzschild presented the first exact solution of Einstein's field equation for massive non-rotating and non-charged spherical objects. His results were astonishing: if the body's density is high enough the deformation of space-time is so high that it can be considered a sort of bottomless pit (Fig. 1.10).

Essentially, Schwarzschild's solution predicted the existence of *black holes* (a name given in the late 60s by John Wheeler), regions of space-time with such an intense gravitational field that nothing can escape to the outside, not even light. There is therefore an imaginary surface that surrounds each black hole (event horizon), characterized by the fact that at each of its points the *escape velocity* equals that of light; once the event horizon is passed, not even light can escape. The escape velocity is the speed necessary to escape a gravitational field and is easily calculated with a minimum of general physics knowledge. In the case of the Earth, the escape velocity is about 11 km/s in the absence of friction and air resistance. The escape velocity depends on the mass of the planet and its radius. Considering the mass of a planet or a star, if it is imposed that the escape velocity is equal to that of light, it can be calculated the radius (known as the Schwarzschild radius) that a celestial object should be in order to transform into a black hole. For the Sun this radius is 3 km and for the Earth it is only 9 mm. This means that to transform into a black hole, all the mass of the Earth should be contained in a sphere with a radius of 9 mm!

In reality, Einstein was not very enthusiastic about Schwarzschild's results as he did not believe much in extreme solutions that implied infinities and,

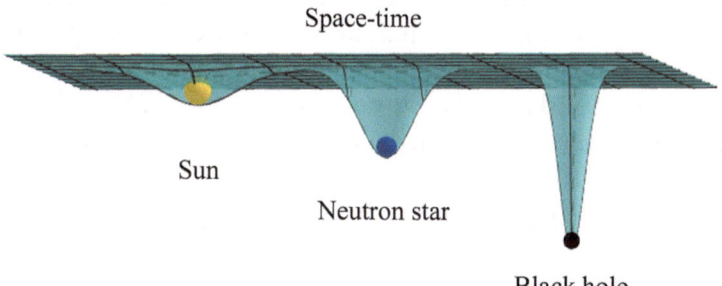

Fig. 1.10 In the case of bodies with very high mass density, the deformation of space-time is such that nothing can escape, not even light. These cosmic objects are called black hole

the black hole was based precisely on a solution of the equations that presented singularities or points where the solutions became infinite.

In reality, these were not physical singularities as they could be eliminated with an appropriate choice of coordinates.

Based on what has been said about gravitational time dilation, the time inside a black hole would be so slowed down as to appear still to an external observer, and vice versa an observer inside the black hole would see events around him proceed at a huge speed. Obviously, it is highly unlikely that a human observer can cross the event horizon to tell his much older great-grandchildren, the incredible experience of seeing objects around him moving at astonishing speeds! In the following years, the concept of a black hole was accepted by most astrophysicists and considered as the final stage of the evolution of a star with a large mass, at least 10 times that of the Sun.

In particular, in 1965 the English mathematician and cosmologist Roger Penrose (Nobel Prize in 2020), showed that General Relativity implied the existence of black holes using very innovative mathematical procedures thanks to which many properties were also understood. Due to their nature and the incredible distance at which black holes were thought to exist, the first direct observation of a black hole only took place in 2019. For the first time a super-massive black hole (M 87) with a mass equal to 6 billion times that of the Sun, 55 million light years away and located at the center of the Virgo A galaxy was photographed.

In 2022, a black hole at the center of our galaxy, the Milky Way, 27,000 light years from Earth and having a mass equal to 4 million times that of the Sun (Fig. 1.11) was photographed. The glowing and luminous ring surrounding the black hole is produced by the light of the photons near the event horizon and whose trajectory is curved by the gravitational attraction of the black hole.

But in addition to being observed, the black hole was also seen in action while devouring a star. In fact, thanks to the sensitive Hubble space telescope, in 2022 astronomers observed an extremely bright event almost 300 million light years away, called *tidal disruption transient event* (*TDE-Tidal Disruption Event*) linked to the extraordinary and fatal encounter of a star with a huge black hole.

Why tidal disruption? Imagine a star as a very large sphere made of gas and plasma that approaches a black hole, the part closest to it feels the force of gravity more than the farther part, giving rise to a tidal effect just like the one observed on the Earth's seas due to the gravitational attraction of the Moon.

In this case, the tidal effect was so strong that the deformation produced was such as to make the star first filamentous and then to literally tear it apart.

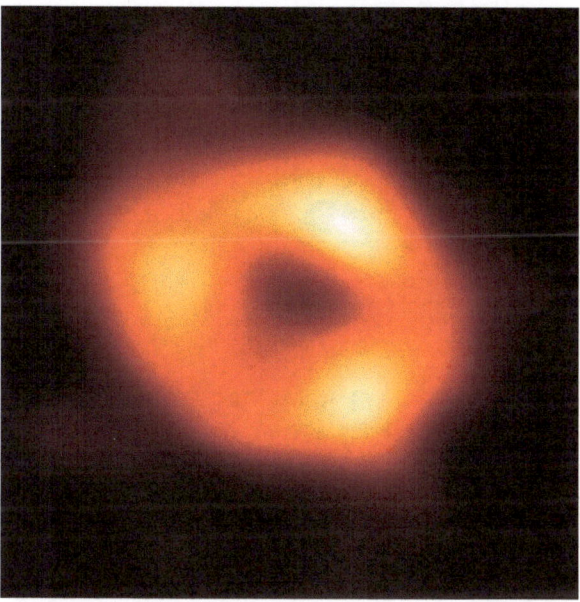

Fig. 1.11 The photo shows the black hole at the center of our galaxy (Milky Way) having a mass equal to 4 million times that of the Sun and 27,000 light years from Earth (credit EHT Collaboration)

The remains of the star wrapped around the black hole like spaghetti around a fork, hence the term *star spaghettification,* thus forming a sort of ball of yarn around the donut-shaped black hole (toroid) with a size equal to the solar system (Fig. 1.12). The succulent stellar donut has gradually been swallowed by the black hole.

The enormous amount of radiation emitted is due to the heating of the stellar gas during swallowing and is measured by the Hubble telescope in the ultraviolet region. The emission of radiation linked to the voracious cosmic meal is mainly in the reason of the X-rays and is emitted by the hot donut formed around the black hole; instead, the emission of ultraviolet rays lasts little and therefore it is not easy to measure it. But thanks to the great sensitivity of the Hubble telescope and the intrinsically very intense event, it was possible to record the ultraviolet radiation, with the significant advantage of obtaining details and valuable information on the dynamics of the cannibalistic cosmic event. These are therefore transient events that can last a few tens of days or a few months and that obviously on a cosmic time scale represent precisely transient events.

The implications of Relativity in the field of astrophysics and cosmology do not end here. Starting from one of the most glaring errors of the great German

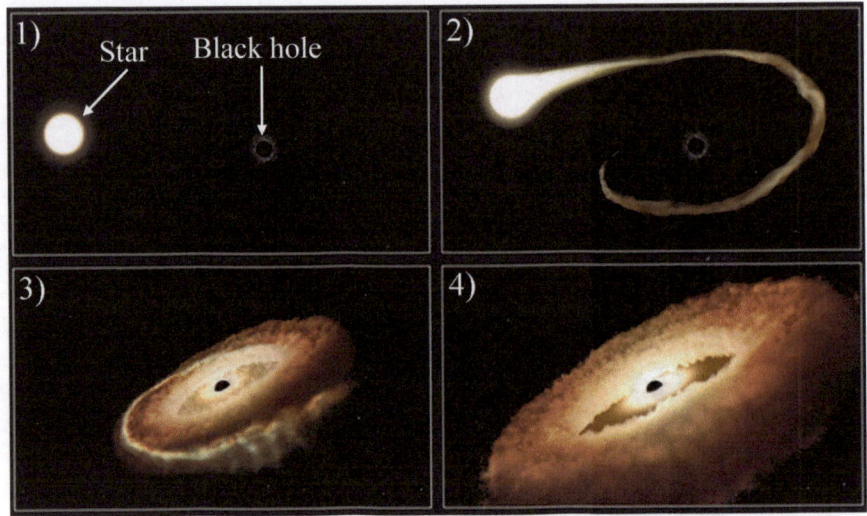

Fig. 1.12 Artistic representation of the various phases of destruction and swallowing of a star by a black hole (credit: NASA, ESA, Leah Hustak STS cl)

scientist, we will see how the most accredited cosmological model is actually a prediction of General Relativity.

Einstein was convinced (and he was not the only one) that the Universe was static, perfectly homogeneous and isotropic, he believed that every direction and every point in the Universe was equivalent to the others. But soon he realized that his vision of the Universe was contradicted by his field equations, which predicted that the masses, driven by the force of attraction, moved and consequently, the same space-time fabric was in continuous movement. To overcome what he considered a problem, in a 1917 article "Cosmological considerations in the general theory of relativity" he introduced into the equations a constant, known as *cosmological constant*, which introduced on a cosmic scale a repulsive compensating force. In 1922 the astronomer Alexander Friedmann, using the theory of General Relativity, derived an equation for the evolution of a homogeneous and isotropic mass distribution known as the Friedmann equation. This equation predicted that the aforementioned mass distribution cannot remain static, but must expand or contract. In 1927 the Belgian physicist and priest George Lemaitre, proposed an expanding Universe. Using Einstein's equations he concluded that the Universe had to expand and predicted a proportionality between the expansion speeds and the distances of the galaxies, which is the basis of the Big Bang model, the cosmological model that predicts a continuously expanding Universe born from a singular event of very high energy. Einstein, while recognizing the correctness of the Belgian astronomer's calculations, did not share the conclusions and addressing Lemaitre he said: "Your calculations are correct, but your physics is abominable". Two years later the astronomer Edwin Hubble, using his telescope in Pasadena, California, verified that Lemaitre's predictions were correct, showing that the galaxies are moving away at a speed proportional to their distance. This extraordinary observation, known as *Hubble's law*, completely contradicted the hypotheses of a static Universe. Faced with experimental evidence, Einstein had to reconsider and in 1933, referring to Lemaitre's theory, he said: "It is the most beautiful and satisfying explanation of creation that I have ever heard" and about the cosmological constant he acknowledged that it was the most and the biggest mistake that he had made in his life. In any case, as Eddington pointed out, the cosmological constant would not have worked because the slightest perturbation would have caused an uncontrolled expansion or collapse. In other words, the cosmological constant described a Universe in precarious balance.

An expanding Universe also solved the well-known Olbers' paradox that has intrigued scientists for about 200 years. In 1826, the German astronomer Heinrich Wilhelm Olbers, proposed the following paradox (which was already being discussed in Newton's time): how is it possible that the night sky is dark? At the time, it was believed that the Universe was static, infinite, eternal, and full of stars. With these assumptions, there should be the same light at night as during the day because, if you imagine being at the center of a series of concentric spherical shells each containing a large number of stars evenly distributed on their own spherical shell, the outer shells should contain many more stars and the dimming of the light due to distance is compensated by the greater number of stars. If the Universe is infinite, it is obvious that the amount of light emitted by the infinite stars is infinitely large, therefore it is not possible that the sky is dark at night. With the Big Bang model that predicts a finite, non-eternal Universe in continuous expansion, the assumptions of Olbers' paradox fall and with them the paradox itself. Our planet is essentially illuminated by the Sun whose intensity clearly surpasses that due to the numerous stars present in our Universe whose estimate is a huge number (over ten thousand billion billion), but not infinite!

Today, the Big Bang model is universally accepted also because it is confirmed by numerous experimental evidences, including the measurement of the *cosmic background radiation* (CBR) carried out in 1964, in a completely random way, by two engineers from the Bell Telephone company, Arno Penzias and Robert Wilson who were making measurements related to the disturbance caused by the Earth's atmosphere in anticipation of the launch of the first telecommunications satellite. The two engineers, now identified as two astronomers, won the Nobel Prize for Physics in 1978 for this important discovery. It is a microwave radiation that pervades the entire Universe and that began to propagate in an isotropic and homogeneous way in space at the end of the so-called dark period of the Universe that lasted about 380,000 years in which, due to the interaction with a very dense gas of electrons and protons, light could not travel freely in space and therefore the Universe appeared dark.

However, an extraordinary discovery made in 1998 by three American physicists, Saul Perlmutter, Brian P. Schmidt and Adam Riess, awarded the Nobel Prize in 2011, seems to bring back into vogue Einstein's cosmological constant. One of the methods to measure the distance of galaxies is to measure the brightness of *supernovae*, that is, massive stars that collapse, whose brightness is so much lower the greater their distance and therefore the distance of the galaxy to which they belong. The aforementioned discovery has shown that the brightness of the *distant supernovae* is 10%–15% less than the value predicted by Hubble's law. This means that the galaxies are further away than expected and therefore they are moving away with a higher expansion speed. Therefore, the introduction of a repulsive force through the cosmological constant whose nature certainly needs to be investigated could explain this extraordinary discovery.

The last topic of this small compendium of Relativity theory concerns gravitational waves, predicted by Einstein in 1916 and directly detected in 2016. They are essentially ripples in space-time that are produced by particularly intense cosmological events. To better understand what they consist of, imagine immersing a stone in a pool and if we start to shake it violently, we will observe that waves are created in the water that propagate along the pool and whose intensity increases the faster we shake the stone in the water. Similarly, if electric charges are oscillated very quickly, electromagnetic waves are generated that propagate in space at the speed of light. The same thing happens in the case of gravitational waves: since masses generate a deformation of space-time, if we imagine accelerating or rotating a mass at high speeds, oscillations of space-time are created that propagate at the speed of light. However, the intensity of such waves is so small that only if we consider large moving masses, such as neutron stars, quasars, or black holes, appreciable gravitational waves are generated that can be measured with extremely sensitive experimental equipment available only today. For this reason, Einstein was very skeptical about their real existence, in fact, twenty years after the first prediction, he dealt with it again in more depth and in 1936 sent together with his collaborator Nathan Rosen an article to the prestigious American journal *Physical Review* titled "Does the gravitational wave exist?" in which he expressed his skepticism, also because in the solution of the field equations they had found points where the solutions became infinite and, as mentioned above, physicists do not like infinities, much less Einstein. In the USA, the method of *peer-review* (peer review) already existed, that is, an article submitted to a scientific journal was sent to experts in the field, on par with the authors, who validated or not its accuracy and scientific reliability. In this case, one of the reviewers had expressed doubts about the method with which Einstein and Rosen had solved the equations. Einstein, not used to this method, got angry and wrote a letter to the editor expressing all his disappointment for having his article read by a so-called expert in the field and withdrew the article. The reviewer, who later turned out to be the great American cosmologist Howard Percy Robertson, contacted Einstein's assistant, Leopold Infeld, and explained to him that in reality the infinite solutions that led to such pessimistic conclusions could be eliminated if a different coordinate system was used. Infeld probably reported to Einstein, who in the meantime had sent the same article to another journal (Journal of Franklin Institute). Realizing the error, Einstein communicated it to the editors of the journal and also changed the name of the article to "On gravitational waves".

As mentioned above, gravitational waves can only be detected on Earth if they are generated by extreme cosmological events, such as those listed below. Gravitational collapse: at the end of its life, a star with high mass explodes (supernova explosion). Rotating neutron stars or pulsars: very dense objects with a mass of about 1.4 times the mass of the Sun and a radius of about 10 km that rotate rapidly emitting electromagnetic radiation and gravitational waves. Binary systems: two pulsars, or a neutron star and a black hole. Coalition and fusion of two black holes: the most extreme event in which two black holes begin to rotate around each other and at a certain point they merge emitting enormous energy also in the form of gravitational waves (Fig. 1.13). But what is the effect of gravitational waves if we were to be hit by them? Since they are oscillations of space-time, at the passage of a gravitational wave there is a tidal effect, that is, a periodic contraction and stretching of distances and therefore also of any object. Depending on the direction of the gravitational wave, we can imagine a person hit by a gravitational wave that periodically becomes taller or shorter or thinner or fatter with the frequency of the gravitational wave. Obviously, let's not worry about seeing such strange things while we walk the streets or have dinner with friends as the contractions and stretches are so small (billionths of the size of a hydrogen atom) that no one would ever notice.

The first indirect observation of gravitational waves was made in 1974 by R.A. Pulse and J. H. Taylor, who were awarded the Nobel Prize in 1993. The

Fig. 1.13 Representation of space-time oscillations (gravitational waves) produced by the coalition of two black holes or neutron stars. Credits: R. Hurt (Caltech-IPAC)

two American scientists studied a system of two neutron stars in orbit, rotating around each other, at a distance of one million km from each other and at a distance of about 1500 light years from Earth.

According to the predictions of General Relativity, the two neutron stars lost energy due to the emission of gravitational waves and therefore should have approached each other by 3 mm for each orbit lasting about 8 h. Consequently, a decrease in the orbital period should have been observed. The measurements of the orbital period made between 1974 and 2004 were in perfect agreement with the theory of General Relativity. But the first direct observation of gravitational waves, in which the incredible tidal effect described above was measured, took place on September 14, 2015 at the LIGO astronomical observatory (acronym for Laser Interferometer Gravitational-Wave Observatory) consisting of two detectors located in Livingston (Louisiana, USA) and Hanford (Washington, USA). At that precise moment, the Earth was crossed by a gravitational wave produced by the coalition of two black holes with masses equal to 36 and 29 times the mass of the Sun and distant 1.3 billion light years from Earth. Downstream of the process of fusion of the two black holes, a mass equal to 3 times the mass of the Sun was transformed into energy, flooding the entire Universe with gravitational waves whose intensity clearly decreased as the distance from the catastrophic event increased.

The two detectors are essentially very precise interferometers similar to those used by Michelson and Morley to verify the existence of the ether

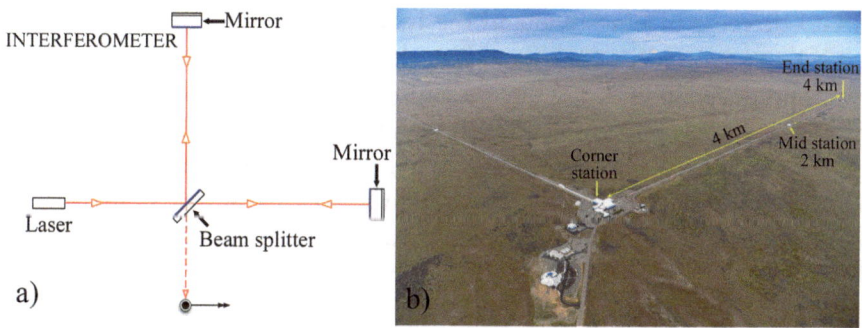

Fig. 1.14 (a) Layout of a basic Michelson interferometer with a laser as a light source. (b) Aerial photograph of LIGO Hanford Observatory showing the scale of the instrument and the locations of the "Corner Station" (where the laser is generated) and one arm's "End Station", where the all-important test-mass mirror resides. Note that the arm is so long that the perspective distorts the distance between the mid and end Station. (Credit: Caltech/MIT/LIGO Lab)

(Fig. 1.14). In a very schematic way, a beam of light is split into two and sent in two perpendicular arms each 4000 m long. At the end of each arm, a particular mirror reflects the light rays, which arriving at the point where they were split, they can give rise to constructive or destructive interference, depending on the path they have traveled. Therefore, by observing the type of interference and more precisely the interference fringes, it is possible to trace back to the difference in the paths taken by the two light rays and, therefore, to the difference in the length of the arms. When a gravitational wave passes, the two arms of the interferometer lengthen and shorten; therefore, a periodic variation in the length of the arms determines an alternation of constructive and destructive interference, producing a signal at the output of the photodetector. Since the estimated variation on the single arm is of the order of 10^{-18} of its length (equal to the diameter of a proton), every single component of the experimental apparatus must be designed and built with great precision. The observation of the same phenomenon in two sites 3000 km apart at the same instant and the perfect agreement between the experimental data and the theoretical predictions based on the equations of gravitational waves, left no doubt about the reliability of the measurement.

The discovery of gravitational waves had a great media resonance and in 2017 the prestigious Nobel Prize was awarded to Kip Thorne, Barry Barish and Rainer Weiss for their decisive contribution to the realization of the LIGO detector and the direct observation of gravitational waves. In the following years, other observations provided further direct evidence of the existence of gravitational waves, in particular, on January 5 and 15, 2020 the

LIGO detector in Lingstone and the Italian detector Virgo, very similar to LIGO and located in Cascina (Pisa), detected two signals of gravitational waves produced in both cases by the fusion of a black hole and a neutron star, occurring about 900 million and 1 billion light years away. The analysis of the first signal showed a mass of the black hole and the neutron star of about 8.9 and 1.9 times the mass of the Sun respectively, while from the second signal it was deduced that the masses of the two celestial objects were about 5.7 and 1.5 times the mass of the Sun, respectively.

The direct discovery of gravitational waves has given rise to a new branch of astronomical investigations known as "gravitational astronomy" in which the tool of investigation is not electromagnetic waves but gravitational waves that are not disturbed by interstellar dust, as happens instead for electromagnetic radiation. Furthermore, by analyzing the gravitational anomalies of the cosmic background it is possible to investigate and study cosmic phenomena related to the first phase of the Universe in which, as already mentioned, the Universe appeared dark as light could not travel freely in space. Gravitational waves, unlike light, have traversed space from the first moments of the Universe's life, providing the opportunity to investigate many aspects of the Big Bang that are still not very clear.

From this brief review on the theory of Relativity and its implications, it is evident how it, in addition to introducing new concepts and paradigms related to time, space and gravity, has had fundamental impact on other sectors of physics such as astrophysics, cosmology, nuclear physics and the physics of elementary particles.

The great German genius, driven by principles of symmetry, was certainly a great unifier and was deeply convinced that the laws of nature should be of a "local" type, that is, every phenomenon, body or more generally a system must be conditioned only by what happens in the immediate vicinity. It is precisely this deeply rooted position of *local realism* that will lead Einstein to take a very skeptical and critical attitude towards the conceptual foundations of quantum mechanics.

The topic we will discuss in the next two chapters is precisely quantum mechanics, which will show us faces of nature even more extravagant and bizarre than the dilated times, contracted lengths, and deformed space-time we encountered in Einstein's theory of Relativity.

2

Quantum Mechanics: The Bizarre Atomic and Subatomic World

In this chapter we will embark on another journey into the world of modern physics, namely that of quantum physics whose foundations and principles are even more counter-intuitive and far from common sense than the theory of Relativity. Starting from 1900, the year in which quantum physics was born, we will go through the most important steps up to the results of the theory of the standard model of particles and fundamental interactions. We have no pretense of being able to communicate clearly and exhaustively the foundations of quantum physics without leaving any doubt or astonishment in the reader. After all, as one of the founding fathers of quantum physics Niels Bohr said: "Those who are not shocked the first time they come across quantum mechanics cannot have understood it".

2.1 A New Physics for the Microscopic World

If the theory of Relativity can be considered a masterpiece of human intellect essentially by the work of one man, quantum mechanics, that is, the theory that describes the world on an atomic and subatomic scale, is a monumental work due to several talents of last century's physics.

It was precisely the experimental streams that Lord Kelvin spoke of (see Sect. 1.1) that gave rise to this extraordinary theory that introduced new paradigms in physics and had a significant technological impact.

One of the most important inventions of the nineteenth century was the incandescent bulb by Thomas Edison in 1878; with the increasing spread of this new type of lighting, the scientific community turned much interest to

© The Author(s), under exclusive license to Springer Nature Switzerland AG 2025
C. Granata, *A Journey into Modern Physics*, https://doi.org/10.1007/978-3-031-77775-2_2

the study of the interaction of electromagnetic radiation (of which light is made) with matter. In particular, the electromagnetic radiation emitted by a heated body, *black body radiation*, aroused much interest. Any body at a temperature above absolute zero (−273.15 °C) emits electromagnetic radiation whose energy and frequency depend on the temperature. It is important to emphasize that the radiation we refer to is not that related to the reflected light that characterizes the color of a body but a radiation produced by the body itself and strongly dependent on the temperature. For example, if we put a metal bar on a flame, when the temperature of the bar reaches around 500 °C, it begins to emit a faint red light that becomes increasingly intense as the temperature increases, if then the temperature is further increased the color will change becoming orange and then eventually yellow if the temperature is sufficiently high. Obviously the bar emits electromagnetic radiation even at temperatures below 500 °C, but we do not see them because the frequencies of the radiation are lower than those of visible light that our eye can perceive, but if we equip ourselves with a thermal camera that can also see infrared radiation we realize the emission of radiation even at lower temperatures.

One way to experimentally study this phenomenon is to use a device first devised by the German physicist Gustav Kirchhoff consisting of an empty box kept at a constant temperature and having a small hole on one of the walls. If you measure the energy of the electromagnetic radiation emitted from the hole and plot it as a function of the wavelength of the radiation or the frequency, a curve is observed that has a shape similar to a bell with a peak corresponding to a certain wavelength λ_{max}. The size (not the shape) of the curve and the value of λ_{max} depend on the temperature. In particular, as the temperature increases the amplitude of the curve increases and the peak shifts towards smaller wavelengths, which is equivalent to saying that as the temperature increases the body becomes more clear and bright, as happens for the metal bar on the flame (Fig. 2.1).

Experimentally it is observed that the product of the temperature T (expressed in Kelvin degrees) and the λ_{max} is equal to a constant; this phenomenon is called Wien's displacement law from the name of the German physicist Wilhelm Wien who investigated the phenomenon and derived the law. Since stars, including the Sun, can be considered black bodies, by measuring the most intense wavelength of radiation (λ_{max}) that comes from a star or the Sun it is possible, using Wien's law, to determine the temperature on the surface of that star. For example, in the case of the Sun the λ_{max} is 500 nm to which corresponds a temperature of 6727 °C.

In the case of a human body at a temperature of 37 °C, the peak corresponds to a wavelength of 9 thousandths of a millimeter (9 μm) and for

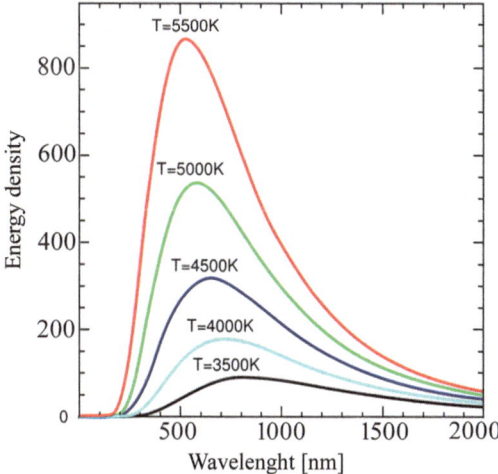

Fig. 2.1 Emission spectrum of a black body at different temperatures. The formula that describes the curves shown in the figure was derived in 1900 by Max Planck who first introduced the concept of energy quantum, giving rise to quantum physics

wavelengths above 3 mm there is no emission. Since these wavelengths fall in the infrared region, the human body is not visible in the dark unless an infrared thermal camera is used.

The cosmos itself can be considered a huge black body, as demonstrated by the measurements made since 1989 by the COBE satellite on the cosmic background radiation from which a perfect black body spectrum with a peak of λ_{max} equal to 2 mm corresponding to a temperature of 3 K (−270 °C) is evident.

Naturally, the cosmic background radiation, an indelible signature of the Big Bang theory, has cooled as the Universe has expanded: it is estimated that at the time of decoupling between matter and radiation (380,000 after the birth of the Universe) the radiation had a black body spectrum, with a peak in the near infrared corresponding to a temperature of about 4000 K, equal to the temperature of the Universe just before the decoupling between matter and electromagnetic radiation.

Another characteristic of a black body or in general of any body at a temperature above absolute zero (−273.15 °C) is the power radiated at a given temperature. Known as Stefan-Boltzmann's law, the power radiated per unit area is proportional, through a constant σ, to the fourth power of the temperature ($P = \sigma T^4$); therefore if the temperature doubles the power emitted increases 16 times. Using Stefan-Boltzmann's law, the power emitted by a human can be estimated: at a body temperature of 36–37 °C, the power is about 100 W, which means that 20 people in a room are equivalent to a 2 kW electric heater.

Using the laws of classical physics (thermodynamics and electromagnetism) it was not possible to explain the trend of electromagnetic energy emitted by a body as a function of wavelength at a fixed temperature. In 1900, the German physicist Max Planck (Nobel Prize for Physics in 1918) proposed a very extravagant solution to explain the phenomenon that he himself did not fully believe in, indeed he considered it a kind of mathematical forcing to explain experimental results. Planck hypothesized that at thermal equilibrium the exchange of energy between electromagnetic radiation and the walls of the black body could only occur in a quantized manner, i.e. for packets of discrete energy and multiples of the quantity $h\nu$ where h is a constant that was later called Planck's constant and ν is the frequency of electromagnetic radiation. Planck's strange hypothesis perfectly explained the experimental data and laid the foundations for a new theory destined to change the way we see the world. The value of Planck's constant is very small and is 6.63×10^{-34} J · s.

In 1905, the young Einstein, using Planck's hypothesis, was able to explain another very interesting phenomenon that could not be understood with the knowledge of classical physics, namely *the photoelectric effect*. This effect involves the emission of electrons from metals when hit by electromagnetic radiation with a frequency greater than a certain threshold that depends on the material. Einstein's theory of the photoelectric effect was reported in one of the three famous articles published in 1905, when the young physicist was employed at the patent office in Bern, in the German scientific journal *Annalen der Physik*, whose chief editor was Max Planck. Einstein hypothesized that light had a corpuscular nature, in addition to its wave nature, that is, it was made of light particles (*photons*, quanta of electromagnetic radiation) whose energy was given by $h\nu$.

It is worth to note that already in the seventeenth century Newton hypothesized a corpuscular nature of light, but according to Newton these were small particles of "luminiferous" matter that propagate in space like tiny balls. Newton's theory, based on the aforementioned hypothesis, was able to explain the reflection and refraction of light. However, following the important interference experiments carried out by the British scientist Thomas Young in 1801, it was realized that light had a wave nature and Newton's corpuscular theory was abandoned.

Einstein resumed Newton's theory but did not hypothesize material particles but rather light particles without mass and preserved the wave nature of light. The wave-particle dualism intuited by Einstein was able to give a comprehensive explanation of the photoelectric effect. It immediately explained the existence of the threshold beyond which the phenomenon was observed: in fact, if energy is proportional to the frequency and to extract an electron from an atom a certain energy is required, it was obvious that photons with frequency lower than the threshold frequency and therefore with an energy lower than the energy of extraction of the electron did not determine any effect. Moreover, the theory also explained the direct proportionality of the energy with which the electrons were expelled with the frequency of the radiation. In fact, if the energy of the photons was higher than that of extraction of the electrons, it was obvious that the remaining part of the photon's energy was transformed into kinetic energy of the photo-emitted electrons. As said, despite Planck being the father of the concept of quanta, he was not fully convinced; in fact, he greeted with skepticism the submission of the article by the unknown young physicist, but being a scientist of great open-mindedness decided to publish the article.

For the theoretical interpretation of the photoelectric effect, Einstein was awarded the Nobel Prize in 1922. Paradoxically, the great German scientist did not receive the prestigious award for the theory of Relativity that had completely overturned the concepts of space, time and gravity, despite the overwhelming experimental evidence of the validity of the theory with the measurement of the deflection of light rays in 1919 by the English astronomer Eddington (see first chapter). In fact, most scientists and in particular the members of the committee responsible for awarding the Nobel Prize were very skeptical and not inclined to accept Einstein's revolutionary ideas. To this must certainly be added the anti-Semitic positions of the German nationalist party, which did not benefit Einstein whose Jewish origins later led him to emigrate to the United States. Particularly adverse to Einstein's ideas was the physicist Philipp Von Lenard (student of Hertz), a very influential personality, who immediately joined the Nazi party and fervently opposed Einstein's scientific activity, considered, with not a few prejudices, wrong and deceptive. Lenard believed that the theory of Relativity led to deceptive and fallacious results and did not miss an opportunity to support, in any public context, the superiority of the "Aryan" physics developed by German physicists compared to the "Jewish" one developed by Jewish physicists of which Einstein was the main exponent. In 1920 during the conference of medical doctors and scientists, held in Bad Nauheim (Germany), he violently attacked Einstein's Relativity using also anti-Semitic arguments. Nevertheless, much to Lenard's disappointment, the scientific community could not avoid recognizing the scientific value of Einstein's research which in 1922 was awarded the prestigious prize. Obviously, these anti-Semitic attitudes, which from 1935 culminated in real racial persecutions, significantly weakened German physics and more generally European physics, in fact many valuable Jewish physicists moved to the United States and most of them were recruited by Robert Oppenheimer for the construction of the atomic bomb within the Manhattan Project.

The prestigious Göttingen school of mathematics itself suffered a significant downsizing. Emblematic of this is the following episode: in 1934 during a dinner, the Nazi Minister of Education asked the great mathematician David Hilbert, the leading exponent of the Göttingen school of mathematics, if mathematics in Göttingen had suffered from the liberation from Jewish influence, Hilbert's reply was terse and lapidary: "Mathematics in Göttingen? There is no mathematics anymore".

It should be reiterated that, beyond anti-Semitic prejudices against Einstein, truly disruptive theories like that of Relativity, are difficult to accept. The very founder of quantum physics, Max Planck, stated: "A new scientific truth does not triumph because its opponents are convinced and see the light, but rather because they eventually die, and in their place a new generation forms to whom the new concepts become familiar".

Returning to our journey into the quantum world, a further and overwhelming proof of the corpuscular nature of light was provided by the American physicist Arthur Compton (Nobel Prize for Physics in 1927) who in 1922 discovered a phenomenon known as the Compton effect in which a photon, following a collision with an electron, loses energy and decreases its frequency. The explanation of this effect required necessarily a corpuscular nature of light; in fact, applying the classical theory of collisions and assuming that the energy of the photon was equal to $h\nu$, it was possible to account with extreme precision for the experimental results.

Faced with this extravagant wave-particle dualism many physicists remained incredulous also because, as mentioned above, Young had demonstrated the wave nature of light in his interference experiments while Einstein's theory predicted a corpuscular vision in some aspects similar to Newton's theory. As we will see shortly, this wave-particle dilemma will also concern matter.

The explanation of experimental phenomena with the hypotheses of quanta does not end here and also deeply involves the structure of the atom. The atomic theory at the beginning of the last century was established and consolidated, no one had any more doubts about the fact that all matter was made up of elementary units and to remove any doubts had contributed the other important article by Einstein in 1905 on Brownian motions, in which starting from the motion of spherical particles in suspension in a suitable solution, atomic quantities such as Avogadro's number were derived. Einstein's predictions were then experimentally verified by the French experimental physicist Jean Baptiste Perrin, sanctioning the triumph of atomic theory.

But the big doubts, instead, were about the structure of the atom whose indivisibility had been questioned by the discovery of the electron (1897) and definitively by the atomic nucleus (1911). But already in 1905, the British physicist Joseph John Thomson author of the discovery of the electron that earned him the Nobel Prize for Physics in 1906, had proposed the first atomic model in which electrons were inserted in a distribution of positive charge, like the raisin inside a *panettone*, hence the name of the *panettone* model (*plum pudding model*). With the discovery of the positive atomic nucleus the Thomson model was replaced by another model.

In 1911, the count and British physicist Ernest Rutherford (already a Nobel Prize for Chemistry in 1908 for the chemistry of radioactive substances) published some interesting results related to an experiment conducted in 1909 from which it emerged that in reality the atom, and therefore all matter, is an empty structure formed by a small central nucleus of positive charge around which the electrons rotate like in a solar system in which the nucleus represents the Sun and the electrons the orbiting planets.

To understand how empty matter is, consider that the size of a hydrogen atom is 0.5×10^{-10} m while that of the nucleus is 10^{-15} m; one can imagine an atom as a football field where the ball in the center is the nucleus and between the edges of the field (electron's orbit) and the ball there is nothing. Rutherford's experiment was conceptually very simple: bombard a target with small projectiles and check how many passed through and how many instead came back. The target was a thin gold foil, while the projectiles were alpha rays made up of two protons and two neutrons, essentially a helium nucleus. Rutherford observed that most of the alpha particles passed undisturbed and only a very small fraction (0.001%) was deflected by a significant angle or even came back. The count therefore deduced that atoms were empty structures with a small nucleus in the center, most of the alpha particles passed far from the nuclei and were not deflected, those that hit the nucleus were deflected by an acute angle, finally those that passed close to the nucleus, due to electrostatic repulsion were deflected by a small angle indicating that the charge of the nucleus must necessarily be positive as the alpha particles are positive.

Rutherford's planetary model had a short life as it was immediately noticed that, since the electron is charged and spinning around the nucleus, it should have radiated energy since, according to classical physics, a charged particle accelerated or decelerated radiates electromagnetic energy. The electrons would therefore have lost all their energy and should have collapsed onto the nucleus in less than a billionth of a second.

Here comes another giant of quantum mechanics, the Danish physicist Niels Bohr who in 1913 proposed a new atomic model introducing quantum discontinuity also in the atom.

First of all, Bohr assumes that all atoms are strictly equal and in the simplest case of the hydrogen atom formed by a proton and an orbiting electron, the distance of the electron from the nucleus is always the same for all hydrogen atoms. Therefore, according to Bohr, it is not possible to think of a planetary model where the orbital radius depends on the initial conditions such as the speed with which the planet begins to rotate around the star. Bohr's other great insight was the quantization of angular momentum which in classical physics is linked to rotations and therefore to orbits. According to the Danish physicist, the angular momentum could assume only values equal to whole multiples of Planck's constant, which was equivalent to saying that the radii of the orbits were quantized and therefore the electrons could only move on certain orbits defined by a precise radius (Fig. 2.2).

This assumption also involved the quantization of the electron's energy, that is, the electron had an energy that depended on an integer n that characterized the orbit and as n increased, the electron's energy decreased as it occupied an outer orbit and therefore was subjected to a lesser force electrostatic.

Another fundamental point of Bohr's model is the following: if an atom was hit by a photon having an energy equal to the difference in energy between two contiguous orbits, the electron jumped into an orbit with a larger radius and absorbed the incident photon, if instead it had spontaneously jumped into an orbit with a smaller radius, it would have emitted photons (Fig. 2.2).

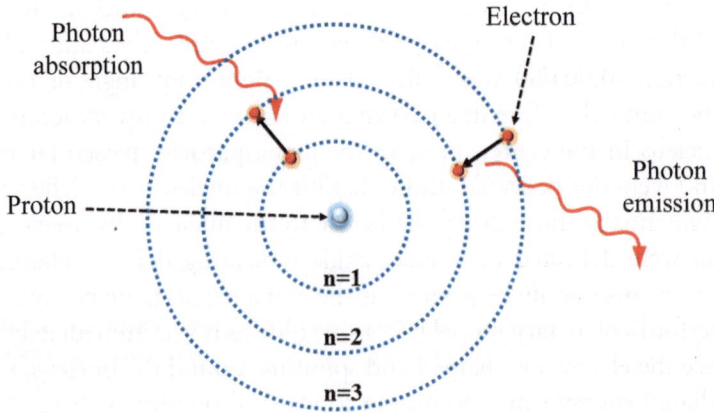

Fig. 2.2 Bohr's atom model. The radii of the orbits and their relative energies are quantized. When an electron jumps from one orbit to another, it emits or absorbs a photon

We are beginning to talk about energy or the difference in energy of an electron between two orbits, but how large are these energies at play? These are energies much smaller than those we encounter in the macroscopic world, so a different unit of measure is used than the one we are used to in the macroscopic world. As is known, energy in the international system of measurement (SI) is measured in Joules (J) which is equivalent to the energy spent to move a body one meter if a force of one Newton is applied. To give a practical example, the energy to lift a one-liter bottle of water by one meter is about 10 J. The energy difference of the electrons in the orbits of a hydrogen atom is on the order of ten billion billion times smaller than a Joule. Therefore, for practical purposes, we use *the electron-Volt (eV)*, 1 eV = 1.6×10^{-19} J which is much smaller than the Joule (1.6 J divided by one followed by 19 zeros!). The name electron-Volt comes from the fact that 1 eV is the energy acquired by an electron if placed in a region of space where there is a potential difference of 1 V.

The great success of Bohr's model was to be able to explain the emission and absorption spectra of atoms. But what are the emission and absorption spectra?

It was already known from the end of the nineteenth century that, when a gas of hydrogen or any other element is excited, for example by heating it, and the light emitted by it is passed through a *optical prism* (optical instrument that separates the various frequencies of light), instead of obtaining the continuous decomposition of light into the typical colors of the rainbow, only very narrow lines of colors on a black background were observed. This means that excited hydrogen atoms emit only certain optical frequencies, called the line emission spectrum. Similarly, if white light passes through a hydrogen gas and then the optical prism, the typical rainbow spectrum is observed with very narrow black lines, indicating that the atomic gas has absorbed some frequencies generating a spectrum known as the absorption spectrum. Bohr's model explained these strange observations completely inexplicable in the context of classical physics. In fact, in excited hydrogen atoms the electrons transition to a higher orbit, but being the latter unstable, they quickly decay to the lower orbit and emit photons whose energy and therefore the frequency is given by the difference in the energy of the orbits divided by Planck's constant, thus explaining the existence in the emission spectrum of discrete lines. A similar argument applies to the absorption spectrum in which only photons having energy equal to the difference in energy of two contiguous orbits are absorbed. For the aforementioned results in 1922 Bohr was awarded the Nobel Prize for physics.

By now there were too many clues to think that Planck's hypothesis was just a mathematical device to explain black body radiation, and so Planck himself, very skeptical, had to convince himself of the strange behavior of nature and regretfully said: "I have tried for many years to save physics from levels discontinuous energy".

But the strangest hypothesis, which is at the base of the development of quantum theory, is that of the French count of Piedmontese origin Louis De Broglie, who in 1924, starting from the wave-particle dualism for photons introduced by Einstein, proposed that matter could also have a wave nature. In other words, a body can behave like a wave, giving rise to typical wave phenomena such as interference and diffraction. In particular, according to De Broglie's hypothesis, the wavelength λ associated with a particle is equal to $\lambda = h/mv$, where m and v are respectively the mass and speed of the particle and h is the usual Planck constant. One of the members of the commission was Paul Langevin who, despite strong skepticism, decided to send the doctoral thesis to Albert Einstein for an authoritative opinion. Einstein was positively struck by it to the point of declaring that De Broglie "had lifted a corner of the great veil". Obviously, beyond Einstein's authoritative judgment, if De Broglie's hypothesis had not given a strong indication for a simple and direct experimental test, it would have been set aside and dismissed as the crazy idea of a wealthy young doctoral student. In 1927, the two experimental physicists Clinton Davisson and Lester Germer fired a beam of electrons at a nickel crystal and observed a diffraction spectrum typical of a wave, confirming De Broglie's revolutionary hypothesis. As the great American physicist Richard Feynman said, "A phenomenon that is impossible to explain classically and that contains the heart of quantum mechanics". The wave nature of matter was subsequently confirmed by further experiments including the famous double-slit experiment first performed in 1974 by three Italian scientists (Pier Giorgio Merli, Gian Franco Missiroli and Giulio Pozzi) in which electrons passed through a double slit one at a time, giving rise to the typical interference pattern of wave phenomena (Fig. 2.3). According to a survey launched by the Journal *Physics World* in 2002, the double-slit experiment using single electrons was the most beautiful physics experiment ever conducted.

Among the physicists of the time, the Hamletian question was: are electrons waves or particles? Niels Bohr proposed the *principle of complementarity*, according to which to interpret the results of an experiment one must adopt the wave point of view or the corpuscular one, but not both descriptions simultaneously. Therefore, an elementary particle can show both wave and corpuscular behavior, and each of these is complementary to the other. Although it is not entirely correct, one can imagine that elementary particles propagate as waves but interact as particles.

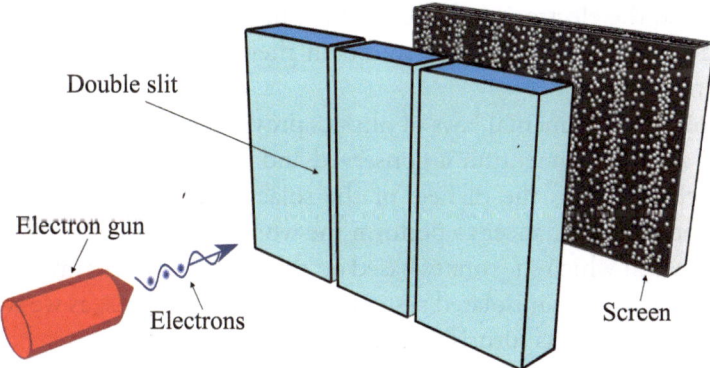

Fig. 2.3 Schematic representation of the double-slit experiment. The dots on the screen represent the incident electrons. Even by sending only one electron at a time, after a large number of electrons (1000), an interference figure typical of wave phenomena takes shape

We observe that macroscopic objects, having a much larger mass than elementary particles, have a practically imperceptible wavelength. In fact, if we calculate the De Broglie wavelength of a grain of sand of just one gram, with a speed of 1 mm/s, we obtain a wavelength less than 10^{-27} m, practically nothing!

But there is an even more extravagant aspect of the wave nature of matter. If you look at the electron at the exit of one of the two slits, it loses its wave nature and behaves like a particle, that is, an interference figure is no longer observed on the screen but only two clear stripes corresponding to the slits. Naturally, the electron is not looked at with the eye but with a detector that interacts with it and according to many scholars it is precisely this interaction or external disturbance that makes the electron lose its wave nature. This phenomenon remains a mystery even today, a mystery that has stimulated various interpretations of which we will speak in the fourth chapter dedicated to the conceptual aspects of quantum mechanics.

A physical quantity typical of the microscopic world and in particular of subatomic particles (electron, proton, neutron, etc.), is the *spin*. It was hypothesized by the Austrian physicist Wolfang Pauli in 1925 to explain both some properties of the elements of the periodic table and some experiments by Otto Stern and Walther Gerlach carried out in 1922 on neutral atoms. It is a sort of intrinsic angular momentum that in classical physics is linked to rotations. One could imagine elementary particles as tiny balls, which can rotate on themselves and depending on the direction in which they rotate we speak of spin up or down. But in reality there is no real rotation associated with the spin; in fact, the electron or the photon are not classical particles, like crystal balls, and therefore they cannot rotate on themselves but are still equipped with spin.

One of the fundamental laws of physics provides that in an isolated system the total angular momentum is conserved and this allows the Earth to rotate indefinitely on itself, the planets of the solar system to follow a flat orbit around the Sun or a dancer to perform the wonderful pirouettes by modulating the speed at which she rotates based on the position of her arms. Therefore, also the total spin of an isolated system is conserved and this, as we will see in the fourth chapter, has surprising consequences.

According to the analogy of rotation, being the electron charged, it becomes a small *magnet* that interacts with the magnetic field; therefore, every charged particle equipped with spin can be imagined as a microscopic magnet or magnetic needle, equipped with a magnetic moment M (physical quantity that quantifies the intensity of a magnet). Also in this case, it is only an analogy that comes from the desire to visualize the quantum world, which unfortunately is not possible. After all, to explain the effects of an electron in a magnetic field, one should hypothesize a rotation speed on itself greater than the speed of light, which is practically impossible. In any case, the magnetic moment of a particle exists and is proportional to the spin through a coefficient, called the gyromagnetic factor. The values that the spin can assume are quantized and can be whole or half-integer. In the case of the electron, the proton and the neutron the values are half-integer: $\frac{1}{2} h$ and $-\frac{1}{2} h$; in the case of the photon, the spin is whole and the possible values are: h and $-h$. Consequently, placed in a magnetic field the electrons or protons align with spin parallel to the magnetic field, just as a compass needle does in the earth's magnetic field. Depending on the value of the spin, particles are classified as *fermions* (half-integer spin) or *bosons* (integer spin). The names fermions and bosons derive from the names of the Italian physicist Enrico Fermi and the Indian physicist Satyendra Bose respectively. It is not a formal classification but deeply substantial as, as we will see, the behavior of bosons and fermions is very different. In this regard, it is necessary introduce one of the fundamental principles of quantum mechanics that plays a fundamental role in the characteristics of atoms, molecules and more generally of all matter: the Pauli exclusion principle thanks to which the Austrian scientist received the Nobel Prize for physics in 1945. Introduced in 1925 to account for some experimental observations related to the ionization potentials of atoms, the Pauli principle asserts that two electrons or more in general two fermions cannot occupy the same quantum state. The immediate consequence of this principle is the occupation of the energy levels of atoms: in the lowest level only two electrons can coexist; one with spin oriented upwards also called *spin up* (½ *h*) and the other with spin oriented downwards called *spin down* (−½ *h*). Therefore, atoms that have more than two electrons are forced to occupy higher energy levels and therefore further from the nucleus, consequently as the number of electrons in an atom increases, so does the size. If the Pauli exclusion principle did not hold, we could have a lead atom in which all 82 electrons occupy the first energy level, with a significant increase in mass density. Moreover, all atoms, having all electrons in the ground state, would behave like noble gases, that is, they would not interact with any atom, not form molecules and more complex compounds. In other words the Universe would look very different from how we observe it and certainly life would never have arisen on our planet without Pauli's principle. We can therefore affirm that at the base of chemistry and therefore also of biology there is a quantum principle that has no correspondence in the world of classical physics. The exclusion principle is not valid for bosons and by virtue of this, as we will see in the section dedicated to condensed matter, some materials exhibit very particular behaviors.

The different behavior of fermions and bosons is even more evident at low temperatures since the quantum aspects prevail over the classical ones. In fact, by decreasing the temperature the thermal agitation of the particles decreases and therefore their speed, leading to an increase in the De Broglie wavelength which is inversely proportional to the speed of the particle.

This last comment introduces us to a very important aspect of the microscopic world: how does temperature affect the behavior of quantum particles? It is worth at this point opening a small parenthesis on quantum statistics that play a fundamental role for the study of sets or aggregates of particles and therefore for the matter that surrounds us and of which we are made.

The link between temperature and speed of the particles of a gas was derived by the German physicist Rudolf Clausius, laying the foundations of the kinetic theory of gases, subsequently formalized by Maxwell and by the Austrian physicist Ludwig Boltzmann, who introduced the statistical aspect through a mathematical function, which allows to calculate the probability that a certain particle of a gas has a certain speed. If this function, known as Maxwell-Boltzmann distribution in function of the speed of the particle, is plotted on a Cartesian plane, a bell shape dependent on temperature is obtained that becomes narrower and more peaked as the temperature decreases with peak values corresponding to the most probable speed. Moreover, as the temperature decreases, the peak shifts towards lower speeds. Therefore, lowering the temperature of a gas of particles is equivalent to decreasing their speed. The great merit of Maxwell and Bolztmann was precisely that of having introduced in the study of thermodynamics the statistical aspect giving rise to the birth of that branch of physics known as *statistical mechanics*. When the speed of the particles is such that their wavelength is large enough to lead to an overlap of the waves associated with them, quantum aspects appear and the Maxwell-Boltzmann distribution is no longer able to explain the behavior of the particles. We are therefore talking about a thermal wavelength which, for sufficiently low temperature values, is comparable to the distance between the particles, causing the overlap of the waves associated with them. An immediate consequence of this is indistinguishability, that is, it is no longer possible to distinguish two identical particles and therefore one of the fundamental assumptions on which the classical Maxwell-Boltzmann statistics was derived falls. At this point, quantum statistics come into play, which depending on the type of particle are distinguished in statistics of *Bose-Einstein* introduced by Satyendranath Bose and Albert Einstein for bosons or *Fermi-Dirac* introduced by Enrico Fermi and Paul Dirac for fermions and play a fundamental role in the study of condensed matter. Quantum distributions allow us to calculate, at a fixed temperature, the average number of particles having a certain energy. In other words, they allow us to determine the effect of temperature on the energy of particles including electrons in matter and photons. Remember that a thermal energy is associated with a temperature, simply given by the product of the temperature expressed in kelvin degrees and the Boltzmann constant $k_B = 1.38 \times 10^{-23}$ J/k, that is $E_T = k_B T$. The mathematical functions that are the basis of the aforementioned statistics have exponential trends; in particular, they decrease rapidly if the energy of the particles is greater than the thermal one, tending towards the classical Maxwell-Boltzmann statistics. This typically happens at high temperatures, since the speed of the particles increases and the corresponding kinetic energy increases more than the thermal energy increases. Also, as already mentioned, by increasing the speed of the particles, the wavelength associated with them decreases and consequently the partial overlap of the relative waves also decreases.

In addition to free particles such as non-interacting atoms or molecules, quantum statistics also apply to electrons in atoms and solids and also to photons. For example, if we have a set of monoatomic hydrogen atoms at room temperature, almost all of the atoms will be in the ground state, since the thermal energy is much lower than that needed to transition an electron from the ground level to the first excited level. The thermal energy at 27 °C (300 K) is equal to one thousandth of eV a value much smaller than the difference between the ground level and the first excited level of the hydrogen atom, equal to about 10 eV. If instead we increase the temperature to several thousand degrees, the electrons distribute themselves in the energy levels following quantum statistics.

In the case of electrons in a metal, already at a temperature just above absolute zero, there will be electrons in states with slightly higher energy, since there is no energy gap. Finally, a set or gas of photons can be described by Bose-Einstein statistics since photons are particles with spin 1 and therefore bosons. Also in this case, if the energy of the photons is much greater than the thermal energy, the exponential term quickly goes to zero indicating that there are no photons with that energy at the considered temperature. For example, in the case of the black body spectrum of the Sun, we have seen that experimentally a bell curve is measured peaked on the frequency of yellow. This implies that the probability of finding a photon with an X-ray or gamma frequency is extremely low. After all, the temperature on the surface of the Sun (6000 K) corresponds to an energy of about 1 eV comparable to that of yellow photons (2 eV). Instead, X and gamma photons have energy greater than thousands of eV. In this regard, we recall that the Bose-Einstein statistics was born in 1920 from the successful attempt by Satyendra Bose to derive Planck's black body law from statistical considerations. Einstein, to whom Bose sent the work, extended the results to atoms as well.

The main difference between the two quantum statistics is that in the case of bosons, a single ground state is predicted in which all particles can exist, while for fermions this is not possible due to Pauli's exclusion principle.

It is not our intention to delve further into quantum statistics to avoid falling into formal technicalities that are not functional to the descriptive and popularizing purpose of the book, however it is important to know that thermal effects on quantum particles are important and are described by appropriate mathematical functions that take into consideration the peculiarities of the particles such as indistinguishability and Pauli's exclusion principle in the case of fermions.

Finally, the other profound break with the concepts of classical physics came from the famous uncertainty principle introduced in 1927 by another founding father of quantum mechanics, the German physicist Werner Heisenberg (Nobel Prize for Physics in 1932), according to which it is not possible to determine with extreme precision and at the same time the position and speed of a particle. This is an intrinsic uncertainty and not related to instrumental errors or to other sources of external error. The greater the precision with which the position is determined, the less is that with which the speed is determined. In particular, the product of the uncertainties cannot be smaller than Planck's constant divided by the mass of the particle m, that is: $\Delta x \, \Delta v \geq h/(m 2\pi)$ where Δx and Δv are the uncertainties on position and speed respectively.

Since Planck's constant is very small, at the macroscopic level the consequences of the uncertainty principle are imperceptible and the precision with which measurements can be made is largely limited by experimental errors (statistical and instrumental).

This principle is in stark contrast with Newtonian mechanics. In fact, according to the Heisenberg principle it is not possible to determine with absolute precision the position and speed at the initial moment of a particle. As a consequence, it does not allow to determine in the atomic and sub-atomic world the trajectory of a particle.

But the uncertainty principle also applies to other physical quantities and in particular to energy and time, that is it is not possible to determine with arbitrary precision the energy involved in a process and the relative time in which it occurs. Therefore, an analogous relationship to the position and speed of a particle can write for the energy and time: $\Delta E \Delta t \geq h/2\pi$, where ΔE and Δt are respectively the uncertainties of energy and of time. This means that for an extremely short time the principle of conservation of energy can be violated as there can be an energy fluctuation or in simpler terms, energy can appear out of nowhere and then disappear in a span of time such that their product is at least equal to Planck's constant. The energy-time uncertainty principle, together with Einstein's famous relation $E = mc^2$, has fundamental implications for the theory of forces or fundamental interactions. As we will see later, some scientists have also hypothesized that at the zero time of the Big Bang there was an energy fluctuation, a consequence of the uncertainty principle, which gave rise to the formation of the Universe.

2.2   The Birth of Quantum Mechanics

Returning to the development of quantum mechanics, the times now they had matured to develop a theory that could coherently and without using ad hoc hypotheses explain the quantum phenomena observed up to that point and predict others. Between 1925 and 1926 two equivalent versions of quantum theory were developed: matrix mechanics and wave mechanics essentially developed by Heisenberg and Austrian physicist Erwin Schrödinger respectively. The latter, based on De Broglie's hypothesis, arrived at the formulation of a wave equation that bears his name (Schrödinger's equation) and which can be considered the fundamental equation of quantum mechanics, analogous to Newton's famous formula for classical mechanics. Like most of the fundamental equations of physics, it is a differential equation whose solution is a wave function (usually indicated with the Greek letter Ψ) which was initially thought to represent an electronic wave, but to which the German physicist Max Born in 1927 gave a probabilistic interpretation: the square modulus of the aforementioned function represented the probability of finding the particle in a given point and at a fixed instant. For his interpretation of the wave function and more generally for his fundamental research in quantum mechanics, Born received the Nobel Prize for Physics in 1954.

It is important to clarify, from the outset, that quantum mechanics is not a probabilistic theory based on the causality of events but a decidedly deterministic theory in which Schrödinger's equation allows to determine with extreme precision the spatial and temporal evolution of the wave function. The probabilistic aspect is only related to the measurement process. This important aspect will be further explored in the fourth chapter, in which the conceptual aspects of quantum mechanics and those related to measurement will be discussed.

Born's interpretation implies that the sum of the square modulus of the wave function at all points in space, must be equal to one, as the probability of finding the particle throughout the space is necessarily equal to one. From a mathematical point of view this is equivalent to saying that the integral of the wave function extended to all space must be equal to *1*.

It is quite intuitive to imagine that, in order for the aforementioned condition, known as normalization of the wave function, to hold, the latter cannot assume increasing values as the distance increases otherwise the sum of the square modulus extended to all space (the integral) would give infinity. Therefore imposing that the wave function is normalizable is equivalent to imposing that it must zero at large distances. This condition plays a very important role in the solution of Schrödinger's equations and its imposition determines in many cases the quantization of the physical quantities under examination.

Another property that the wave function must satisfy is that it must be single-valued i.e. it cannot assume two values at the same point in space. This other condition also leads, in some cases, to quantization, as in the case of a circular motion of a particle in which obviously the value of the wave function cannot change at the same point on the circumference where the particle, after a complete revolution, finds itself.

In general, the quantization of energy, but also of angular momentum or other physical quantities, arises when a particle is forced to be confined in a limited region of space, such as a particle in a box, or in an atom where one or more electrons orbit around a nucleus. The quantization of these states, called *bound states*, is due to the mathematical conditions that we impose on the wave function and therefore ultimately to the wave nature of matter.

A direct consequence of the meaning of the wave function is one of the most surprising effects of quantum mechanics, i.e. *the tunnel effect*. If we suppose to throw a tennis ball against a wall, we expect the ball to always bounce back. However, according to quantum mechanics there is a non-zero probability that the ball will pass through the wall, as if a tunnel momentarily opened in the wall to allow the passage of the ball, just like what happens on platform $9^{3/4}$ at King's Cross station where Harry Potter and the other wizards pass through the wall with their trolleys. However, in the macroscopic world the probability that this happens is so small that it makes the observation of such a phenomenon impossible. But if you go down to the atomic and sub-atomic scale, the probability that a particle will pass through a classically forbidden barrier greatly increases, making this phenomenon really observable. The explanation of this phenomenon comes from the meaning of the wave function whose square modulus, as mentioned, provides the probability of finding the particle at a given point and at a given instant. If you solve Schrödinger's equation by setting the appropriate wave function matching conditions (absence of discontinuity) at the barrier boundary points, you find that the wave function decreases exponentially within the barrier, assuming a non-zero value beyond it. Therefore, there is a non-zero probability of finding the particle on the other side of the barrier, as long as it is not very large.

The tunnel effect has helped to explain some fundamental phenomena of physics such as radioactive phenomena related to *alpha decay* i.e., the expulsion by some unstable nuclei (usually having atomic mass of the order of 200 times the mass of the proton) of alpha particles. In fact, the phenomenon is schematized by imagining that the parent nucleus already contains an alpha particle with a certain energy. This particle, in order to leave the nucleus, must overcome an energy barrier due to the concomitant effect of the repulsive Coulomb forces and the attractive nuclear forces. Since the energy possessed by the alpha particle is always less than the aforementioned potential barrier, the only way for it to leave the nucleus is the tunnel effect. The same mechanism is at the basis of nuclear fusion reactions in which the electrostatic repulsion would prevent, without tunnel effect, the approach of atomic nuclei to a distance such as to make the strong nuclear force predominant over that of Coulomb.

Still within the field of basic physics, the tunnel effect has represented in recent decades a powerful tool for the study of macroscopic quantum effects, helping to clarify important controversies existing since the birth of quantum mechanics such as its validity in the macroscopic world. Moreover, as we will see in the next chapter, this strange effect, completely incomprehensible from a classical physics point of view, is at the basis of important applications ranging from electronics to advanced microscopy.

Another consequence of the meaning of the wave function and the principle of indeterminacy is the concept of *atomic orbital* (Fig. 2.4).

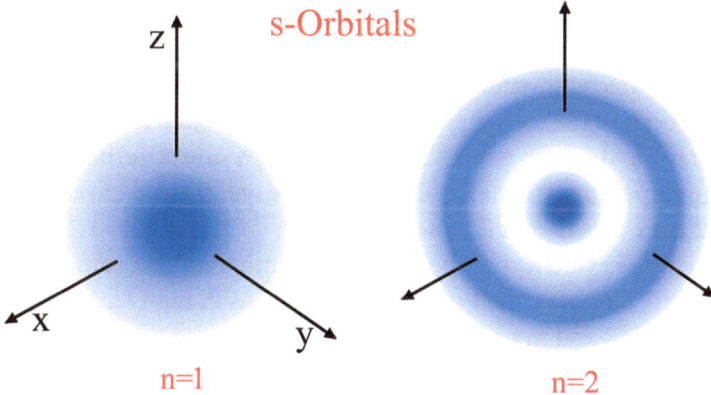

Fig. 2.4 Orbitals related to the first two principal quantum numbers. They represent the region of space where we have a 90% chance of finding the electron. The intensity of the color is proportional to the probability of finding the electron, the white color corresponds to a nearly zero probability

It no longer makes sense to talk about an orbit as a trajectory followed by electrons, but rather of an orbital as a zone of space around the nucleus where the probability of finding an electron is 90%. Therefore, the atom must be imagined as a central nucleus and an electronic cloud more or less dense depending on the value of the square modulus of the wave function. However, it is not a question of an electronic charge delocalized in the region of the orbital, but of a probability cloud in the sense that each point of the orbital is associated with a probability of finding the electron. In the particular case of the hydrogen atom, the resolution of Schrödinger's equation with the conditions on the wave functions mentioned above, allows to determine three integer quantum numbers known as n which is linked to the energy of the orbital and can take positive integer values starting from 1, l which can take values between 0 and $n-1$ and it is related to the shape of the orbital (shell, double drop, etc.) and finally m which is related to the orientation of the orbital in space and can take integer values between $-l$ and l. Finally, as we have seen, there is the spin which in the case of electrons only takes two values, 1/2 and −1/2.

These results, at least from a qualitative point of view, can be extended to all atoms including those containing many electrons. The four quantum numbers together with the Pauli principle allow us to determine the electronic configuration of all elements, that is how the electrons are arranged in the various orbitals starting from the one closest to the nucleus where n is equal to 1 and l is equal to zero (spherical orbital). For example, hydrogen has only one electron that occupies the s orbital at the lowest energy or 1 s, where the number preceding the letter is the principal quantum number, while helium has two electrons that continue to stay in the *1 s* orbital, one with spin up and the other with spin down. Moving on to lithium which has 3 electrons, two occupy the level *1 s* and the third electron due to the Pauli principle cannot stay in the first orbital and will occupy the *2s* orbital with higher energy, for beryllium which has 4 electrons we will instead have two electrons in *1s* and the other two in *2 s* and so on.

It is therefore evident that the quantum numbers with a somewhat magical flavor that are taught in the first chemistry lessons in high schools, come from quantum physics and in particular from the solution of the Schrödinger equation for the hydrogen atom.

In addition to the hydrogen atom, another simple system that plays a fundamental role both in classical physics and in quantum physics is the harmonic oscillator, i.e., a massive body connected to a spring. It is easy to imagine the motion of such a system, at least from a point of view classical: if we pull or compress the body, we observe an oscillation of the body that tends to dampen until it stops due to friction and air resistance. But in the absence of these braking forces the body goes back and forth indefinitely in a manner completely analogous to a pendulum.

Why is it so important to study elastic systems like the harmonic oscillator? Because in reality, even if we do not realize it, molecules, atoms, and nuclei present in all the matter that surrounds us, including those of living beings, vibrate around equilibrium positions. Take the case of water molecules, made up of two hydrogen atoms and one oxygen atom with the well-known *V*-shape in which the oxygen is at the vertex while the two hydrogen atoms are at the ends of the oblique fins that can approach or move away from each other or contract and expand. In both cases, there is a vibration/oscillation of the atoms, just as it happens for the nuclei in a sodium crystal or any solid. These are very high frequency oscillations, for example in the case of water, the molecules flap their wings billions of times per second. Naturally, these atomic or molecular vibrations depend on the temperature and tend to attenuate as the temperature decreases. It is therefore clear that the quantum study of vibrating microscopic objects is very important and since all vibrations can be traced back to harmonic oscillators, it is of fundamental importance to study these very simple mechanical systems.

If you solve the Schrödinger equation for a harmonic oscillator, you find that the energies it can assume are quantized and in particular are half-integer multiples of Planck's constant multiplied by the frequency ν, which depends on the mass of the body and the elastic constant of the spring ($E_n = (n + 1/2)h\nu$).

Unlike the classical case where a harmonic oscillator can have an energy equal to zero corresponding to the case where the body is still, in the quantum case the energy is always different from zero. In fact, for $n = 0$ the energy is $h\nu/2$ and is called zero point energy, this means that stationary quantum oscillators do not exist. After all, if this were possible, it would violate the uncertainty principle, as the condition of rest would be equivalent to a state where the speed and position of the body are exactly equal to zero, a condition obviously not allowed by the uncertainty principle. As for the hydrogen atom, the values of n greater than zero correspond to the excited states that can be obtained by providing energy from the outside. Since the frequency of the oscillator is fixed by the mass of the particle and by the elastic constant of the spring, the excited states are characterized by a greater amplitude of oscillation. Also the thermal energy, if it is sufficient to stimulate the transitions, can induce excitations of the quantum harmonic oscillator. In the particular case of water molecules, at room temperature they are all in the fundamental vibrational state ($n = 0$), as the thermal energy is not sufficient to transition the molecule to the first excited level. But if we use photons with frequency in the infrared field, we can induce transitions to higher vibrational levels just as happens for the absorption spectrum of the hydrogen atom for visible or ultraviolet light. However, there are oscillating systems like the atomic nuclei of a metal for which even a relatively low temperature is sufficient to transition them to the excited states, in this case they distribute themselves in the various energy levels following the predictions from quantum statistics (Bose-Einstein or Fermi-Dirac).

Therefore, in the microscopic world of animate or inanimate matter, there is no state of rest, its elementary constituents, oscillate, vibrate, rotate and translate even if we freeze everything to absolute zero. It is the laws of quantum mechanics that prevent bodies from being still. As we will see in the next paragraphs, these "quantum springs" are at the basis of the quantum theories of fundamental forces and condensed matter.

The new mechanics, based on the Schrödinger equation and reasonable mathematical conditions to impose on the wave function, managed to predict with remarkable precision and without any ad hoc assumptions the behavior of atoms, molecules, condensed matter and in its relativistic version, also the behavior of atomic nuclei, elementary particles and fundamental interactions. Moreover, as we will see later, it has led to a technological revolution of enormous impact (first revolution quantum). Just think of the invention of semiconductor transistors, lasers, the tunneling microscope, imaging diagnostics, and nuclear medicine to realize the technological impact of quantum physics.

2.3 Quantum Mechanics Meets Special Relativity

Schrödinger's equation was formulated essentially from the well-known formula of mechanical energy in classical physics, that is, the sum of kinetic energy and potential energy. But as mentioned in the first chapter, the theory of special Relativity predicts a more general formula for energy in which the energy of mass is also present. Therefore, it was immediately clear that it would be necessary to develop a relativistic quantum theory that included the concepts of the theory of special Relativity. In other words, it was necessary to write a more general equation than Schrödinger's, including cases where relativistic effects could not be neglected, i.e., when the speed is not negligible compared to that of light, a circumstance very common in the subatomic world. The first attempts made in 1927 by the German physicist Walter Gordon and the Swedish physicist Oskar Klein were unsuccessful since the equation they formulated predicted solutions with negative energy difficult to interpret and above all predicted negative probabilities to which, even if one wanted to, it was impossible to attribute a reasonable interpretation.

Here comes one of the most emblematic and important characters of quantum physics, the English theoretical physicist Paul Adrien Maurice Dirac.

Of a shy nature, Dirac was known to be extremely taciturn to the point that his colleagues in Cambridge had ironically established "the dirac" as a unit of measure of talkativeness: one dirac was equivalent to the emission of one word per hour!

The great English physicist in 1928 published an article titled "Quantum Theory of the Electron" destined to become a milestone in quantum physics and in which one of the most beautiful and important equations of physics, known as Dirac's equation, was reported.

The adjective beautiful is not used by chance. Dirac, considered an aesthete of physics, was convinced that "a physical equation should have mathematical beauty"; a phrase he wrote during a visit to the Moscow Academy of Science in 1956 on the famous blackboard on which illustrious scientists were invited to write a phrase representative of their research and which would be left to posterity. On another occasion, perhaps inspired by Galileo (*Il saggiatore 1623*) according to whom "the book of nature is written in the language of mathematics", he stated that "God is a mathematician of the highest caliber and used extremely advanced mathematics to build the Universe". Obviously, beauty is not enough, a physical law must have an indisputable predictive ability verified by experiments. In the case of Dirac's equation, both requests were excellently met.

Dirac's equation solved the problem of the nonsensical negative probabilities of the Klein-Gordon equation, but it continued to predict solutions with negative energy. In fact, the solutions were four: two with positive energy and the other two with negative energy. The two positive energy solutions are associated with the two spin states of the electron, which at this point were not introduced ad hoc as Pauli did a few years earlier, but naturally predicted by Dirac's theory.

But the English scientist went much further and did so with his brilliant interpretation of solutions with negative energies. In 1930, he intuited that those solutions corresponded to particles identical to electrons except for the charge, that is, they had to have a charge equal to that of the electron but positive, in practice a positive electron.

In particular, his interpretation, known as *Dirac's sea*, is based on Pauli's exclusion principle. The minimum possible positive energy for a particle is that of mass, i.e. mc^2 corresponds to the stationary particle.

The spectrum of positive energy states describes particles that move at an energy equal to or greater than mc^2. Below the positive energy spectrum would be the spectrum of negative energy states (Fig. 2.5) with energies starting from $E = -mc^2$ and lower (*Dirac sea*).

At this point, it is legitimate to ask the following question: why doesn't a resting positive energy electron, that is, having the minimum possible positive energy ($E = mc^2$) not indefinitely decay towards the negative energy states that are below it? To answer this question, Dirac assumed that in the vacuum all energy levels were completely filled by particles having the same charge as the electron but negative energies. The electron could therefore not decay into the underlying states due to the Pauli exclusion principle. The vacuum would

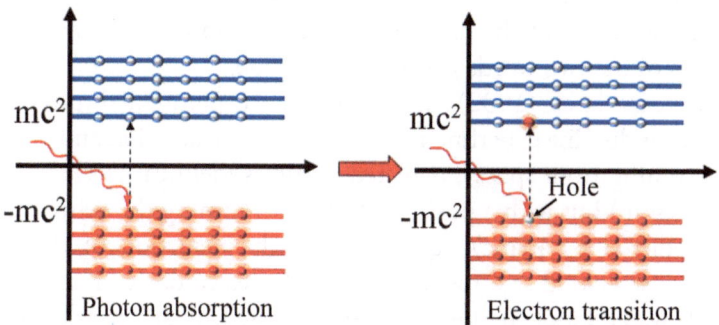

Fig. 2.5 Representation of the Dirac sea. The positron is essentially the positive gap left in the Dirac sea following the transition of an electron from a negative energy level to a empty positive energy level

therefore actually be *full* of a sea of particles. The existence of such sea particles can be experienced, only when a photon, absorbed by one of these negative energy particles, provides enough energy to make it transition into the positive energy region, leaving in its place a *hole* with opposite charge and energy. Indeed, if a minimum energy photon $E_g = 2m_ec^2$ hits a $-m_ec^2$ energy state of the Dirac sea, this, absorbing the photon, will acquire energy m_ec^2 and can therefore transition into the positive energy spectrum part. In the Dirac sea, there will therefore be a hole left by the particle that has transited above and with simple reasoning it can be shown that this hole will have positive energy and a positive charge. Indeed, when a negative energy electron $(-m_ec^2)$ is subtracted from the vacuum, its energy increases by a positive amount: $E'_{vacuum} = E_{vacuum} - (-m_ec^2) = E_{vacuum} + m_ec^2$; since the physically observable quantity is only the energy change between before and after, a change in energy can be measured: $\Delta E_{vacuum} = E'_{vacuum} - E_{vacuum} = m_ec^2$. For the charge, the reasoning is analogous: when an electron of negative energy with charge $(-e)$ is subtracted from the vacuum, its final charge will be equal to $Q'_{vacuum} = Q_{vacuum} - (-e) = Q_{vacuum} + e$. Once again, the physically observable quantity is the change in charge, that is $\Delta Q_{vacuum} = Q'_{vacuum} - Q_{vacuum} = e$. The same applies to spin, the hole will have a spin equal and opposite to that of the negative energy electron.

Therefore, the hole behaves in all respects like a particle of positive energy, positive charge and the same spin as the electron but in the opposite direction and is called an *antielectron*. Naturally, this suggestive interpretation fails for bosons for which the Pauli principle does not apply, but antiparticles also exist for bosons. Moreover, Dirac's sea has never been experimentally observed, so Dirac's original interpretation was later replaced by a more formal explanation within the subsequent evolution of quantum mechanics, namely quantum field theory. With his equation, Dirac had predicted the existence of antimatter and in particular of the antielectron better known as *positron*, from the contraction of the two words *positive* and *electron*.

A few years later (1932), the American experimental physicist Carl Anderson, conducting experiments on cosmic rays (radiation and high-energy particles coming from space), detected particles identical to electrons (same mass, same spin) but with a positive charge, confirming the existence of antimatter predicted by Dirac's equation. For the detection of the positron, Anderson used a cloud chamber consisting of a sealed box containing air saturated with water vapor. If the chamber is crossed by an electrically charged particle, a trail is formed due to the condensation of water vapor around the ionized atoms formed following collisions with the incident charged particle. Anderson applied to the chamber magnetic fields produced by two powerful electromagnets capable of producing a magnetic field of 1.7 Tesla (about 34,000 times the Earth's magnetic field) in order to curve the electric particles that crossed the cloud chamber according to a known law of electromagnetism. By analyzing the trail left by the particle and considering that the radius of curvature depends on the charge and mass of the particle, it was possible to trace its nature.

Anderson's experiment had highlighted the fundamental role of Dirac's equation and its predictive power, in relation to which Dirac himself stated: "the equation was smarter than me". Dirac was awarded the Nobel Prize for Physics in 1933 together with Schrödinger for his fundamental contributions to quantum physics.

Antimatter has a peculiar characteristic: if it comes into contact with matter it annihilates itself, transforming into electromagnetic energy according to Einstein's formula $E = mc^2$.

The prediction and discovery of antimatter represented a huge turning point also from a conceptual point of view as it made us understand that the form of matter we know is not the only one existing but there is and it is possible to produce a form of matter identical to the ordinary one but with the individual constituents (protons and electrons) having opposite electric charge.

Obviously, the fact that there is a Universe made of matter leads us to deduce that the amount of antimatter is negligible otherwise all the matter and antimatter of the Universe would transform into energy through a mega annihilation and then perhaps the energy would convert back into matter, continuing ad libitum this process of annihilation and reconversion into matter. During the primordial period of the Universe, the amount of matter and antimatter was the same and indeed processes of annihilation and transformation into matter alternated in a sort of unstable equilibrium. Based on studies done on cosmic background radiation, it was understood that due to a very slight asymmetry between matter and antimatter, the latter has almost completely disappeared and only matter has survived. We are talking in any case of an extinction of great proportions, it is estimated that only one part of matter out of a billion survived, giving rise, in the billions of years that followed, to galaxies, stars, planets and life.

The small amounts of antimatter existing in nature come from cosmic rays, but these are really small amounts. Even in some foods rich in potassium such as oranges and bananas, positrons are produced from the decay of a potassium isotope (potassium 40), present in nature with a very low percentage, about 0.012%. Remember that the number following the element refers to the atomic mass number, that is the sum of the protons and neutrons inside the nucleus and that the *isotopes* are atoms having the same number of protons and a different number of neutrons. For example, the isotopes of potassium are three: potassium 39 with 19 protons and 20 neutrons, present in nature with a percentage of 93.2%; potassium 41 with 22 neutrons and a percentage of 6.8% and finally the radioactive one, that is potassium 40 with 21 neutrons.

Our own body, containing potassium, emits a very small amount of antimatter in the form of positrons; it is about 170 positrons every hour. In any case, positrons have a very short life, they annihilate as soon as they meet an electron, that is almost instantaneously releasing energy in the form of gamma rays equal to the mass of the electron-positron pairs times the speed of light squared. Since the mass of positrons and electrons is very small (9.2×10^{-31} kg), the energy produced by their transformation into energy is equally small. We can estimate that in an hour our body emits an amount of energy due to the annihilation of the positrons we produce equal to about 100 billion times smaller than a calorie, practically imperceptible.

To produce instead antimatter artificially and in less modest quantities, high-energy photons (gamma rays) are made to collide with matter or large particle accelerators are used in which electrons or protons are made to collide at speeds close to the speed of light. Following the impact, part of the energy transforms into mass giving life to particle-antiparticle pairs. In this way, in 1955 the antiproton was discovered at the particle accelerator of the Lawrence Berkeley National Laboratory in California, the antineutron the following year in the same laboratory. The experiments that led to the discovery of the antiproton were led by the Italian physicist Emilio Segrè and by the American physicist Owen Chamberlain who, received for this discovery the Nobel Prize in 1959. In 1977, at CERN in Geneva about 50,000 anti-hydrogen atoms were created consisting of a negatively charged antiproton and a positively charged positron, while in 2011, at the American national laboratories of Brookhaven the largest nucleus of antimatter was created, namely the anti-nucleus of helium 4, consisting of two antiprotons and two antineutrons.

If matter and antimatter transform entirely into energy, it is legitimate to ask whether it would be possible to exploit the annihilation of matter with antimatter to produce energy. It would be clean energy without any kind of radioactive waste and with an efficiency far superior to all other forms of energy. The annihilation of a kg of antimatter with one of matter would produce about 2 billion times the energy produced by the the combustion of a kg of oil produces and about 70 times the energy produced by the nuclear fusion of a kg of hydrogen. Would it therefore be possible to create engines that use antimatter as in the well-known Star Trek television series? In fact, from a physics point of view, the reasoning is flawless and the thing would be possible. However, in addition to the technological difficulty due to the storage of antimatter, which must be kept away from matter for obvious reasons, producing antimatter is extremely expensive, perhaps the most expensive material in the world. It has been estimated that the cost of a single gram of antimatter is on the order of about 50 billion dollars, which makes it impossible to use antimatter as a source of energy, at least for the moment.

The predictive potential of Dirac's equation does not end with spin and antimatter, but also naturally explains the *fine structure* of the energy levels of hydrogen atoms. An explanation already obtained previously by adding ad hoc additional terms in Schrödinger's equation to account for relativistic effects and the interaction between the electron's spin and orbit. However, a small difference between the energy levels related to the *2s* and *2p* orbitals, measured precisely by Willis Lamb and Robert Retherford in 1947 with a brilliant experiment and known as *Lamb shift*, remained unexplained.

Furthermore, Dirac's equation also manages to predict the gyromagnetic ratio of the electron, which was a real puzzle for physicists of the time. As we saw in the previous paragraph, the magnetic moment of the electron is proportional to its spin through a coefficient called the gyromagnetic ratio. This coefficient, within the framework of classical theory, assumes a value equal to one; instead, experimentally a value was found to be about double that predicted. Solving Dirac's equation for the electron in an electromagnetic field, it is found that the gyromagnetic factor is equal to 2, compared to the first accurate experimental value measured in 1947 by American physicists Polykarp Kusch and Henry Foley, which differs by 0.12% from that predicted by Dirac's equation. For this accurate measurement, Kusch received the Nobel Prize in Physics in 1955. Such a small difference seems insignificant, as do those small separations of the energy levels of the spectra of the hydrogen atom (*Lamb shift*), but they were crucial for the birth of what is considered the most precise physical theory existing from the predictive point of view, i.e., quantum electrodynamics better known as QED (Quantum Electro Dynamics). Considered a real jewel of physics, QED deals with electromagnetic phenomena from a quantum point of view.

There is no doubt that Einstein's field equations (see Chap. 1) and Dirac's equation are the most important in modern physics. As done for the field equation, it would therefore be worth writing and commenting on Dirac's equation, but it is not our intention to make another exception and fail to keep the premise of not using complicated equations or formulas. However, in Fig. 2.6 you can admire an unusual form of the equation, engraved on the commemorative plaque of the great English scientist inaugurated on November 13, 1995 in a nave of Westminster Abbey (London) not far from

Fig. 2.6 Commemorative plaque of P.A.M Dirac on which his famous equation is reported. Inaugurated in 1995, the plaque is located in a nave of Westminster Abbey

the monument dedicated to Newton and from the tombs of other famous British scientists and characters.

The meeting between quantum mechanics and special Relativity was certainly fruitful, leading to the formulation of a more general theory thanks to which it was possible to predict a new form of matter and explain new phenomena.

The same cannot be said of the meeting between the theory of general Relativity and quantum mechanics, which more than 100 years, despite the efforts made by great scientists and primarily by Einstein, has not yet led to a more general theory in which gravity and quantum phenomena can naturally take shape.

2.4 The Infinitely Small and the Fundamental Forces

Dirac does not limit himself to writing one of the most beautiful and important equations of Physics, but goes further and formulates together with Pauli and Heisenberg the first version of the quantum field theory, a further evolution of quantum mechanics of fundamental importance for particle physics and fundamental interactions.

The quantum field theory introduces a new paradigm in the way of conceiving the fundamental forces or interactions and the fields associated with them. We remember that in addition to the well-known force of gravity and the electromagnetic force responsible for interactions between electric charges and therefore also for the chemical properties of the elements, there are two other forces: the strong nuclear and the weak nuclear. After all, if the atomic nucleus is made up of protons which, despite having the same positive charge, do not repel each other, there must necessarily be a force between the protons stronger than the Coulomb force that would tend to repel them. This force is the strong nuclear force, it acts between protons and neutrons, it is always attractive but has a very small range of action, comparable to the dimensions of an atomic nucleus (10^{-15} m) with an intensity that is about 100 times more large than the electromagnetic force and well 10^{38} times greater than the gravitational force. However, not all particles are subject to this force and this allows for another important classification of elementary particles: the *hadrons* on which all forces act like protons and neutrons and the *leptons* on which the strong nuclear force does not act like the electron and the positron.

The weak nuclear force is instead responsible for radioactive processes, that is, the transformation of an unstable atom into another stable one through radiation emission. As an example, we report the one mentioned in the previous paragraph about the natural emission of antimatter: potassium 40 contained in many foods and therefore also in our body, transforms in a very low percentage (0.001%) into argon 40 through radioactive decay of type β^+ in which a proton transforms into a neutron emitting an antielectron and an extremely light particle called *neutrino*. This process occurs thanks to the existence of the weak nuclear force whose range of action is even more small (10^{-18} m) and whose intensity is about 1 million of times smaller than the strong nuclear force but still much more intense than the gravitational force which is by far the least intense force compared to the others. Before resuming the discussion on quantum theory, it is worth opening a parenthesis on the neutrino, particle of fundamental importance also in fields of physics different from that of elementary particles and fundamental interactions.

The neutrino was introduced by Pauli in 1930 to explain some anomalies in the beta decay process in which a neutron is transforms into a proton and an electron. This happens in some radioactive atoms whose nucleus has an excess of neutrons compared to protons. These atoms are not stable and to become so at least one neutron must transform into a proton as happens in the case of cobalt 60, which transforms into nickel 60, emitting an electron also known as *beta ray*. Even the isolated neutron is not stable and in about 18 minutes it transforms into a proton and an electron. In this process the accounts did not add up: the sum of the energy of the proton and the electron was less than that of the neutron, moreover the momentum was not conserved as the particles generated (electron and proton) instead of going in opposite directions, formed an angle between them. All this suggested the existence of a new particle necessarily neutral and very light to which Fermi gave the name of neutrino, imagining a small neutron. Because of the very weak interaction with matter and its very small mass, the neutrino was only discovered in 1956 by American physicists Frederich Reines and Clyde Cowan who used a huge detector mounted near a nuclear power station that produces many neutrinos. Unlike other radioactive waste from nuclear power stations, neutrinos are not at all dangerous as they pass through the human body almost at the speed of light without interacting with any tissue or cell of our body. After all, our body is crossed on average every second by over 400 trillion neutrinos coming from the Sun, significantly higher than those due to nuclear power stations which are at most 100 billion per second or 4000 times less than the solar ones. Moreover, as we have already had the opportunity to say in the previous paragraph, our own body produces antimatter through the decay of potassium 40 and together with antimatter, neutrinos are also produced which are around 4000 per day. The mass of the neutrino has been for many years the subject of study and research. Initially it was hypothesized that the neutrino had zero mass, but subsequent experiments, including the recent one *Katrin* (Karlsruhe Tritium Neutrino) have set an upper limit to the mass of the neutrino which is equal to 0.8 eV/c^2 which is equivalent to saying that the neutrino has at most a mass 630,000 times smaller than that of the electron which is the smallest stable particle in absolute terms. Note that, using the famous equation $E = mc^2$ which we talked about in the first chapter, the mass has been expressed as the ratio between an energy (eV) and the speed of light squared (c^2). Often the term c^2 is omitted and simply the mass of particles in eV and its multiples ($keV = 10^3 eV, MeV = 10^6 eV, GeV = 10^9 eV$). For example, we can say that the electron has a mass of 511 keV, the proton and the neutron of about 1 GeV.

But why is the neutrino so important? This elusive particle was one of the main candidates to explain the famous dark matter that permeates our Universe but does not interact and is not visible. Today, considering its very small mass, there is much skepticism about this hypothesis.

Then there is the problem of solar neutrinos: the flow of neutrinos coming from the Sun is significantly lower than that predicted by the theory of the solar model, challenging the validity of the model. After over 40 years, the problem was solved in 2002 thanks to an in-depth investigation into the nature of neutrinos that led to the discovery of neutrino oscillations. After being produced by the Sun, these tiny particles can transform into neutrinos of another family, thus reducing the number of neutrinos measured by detectors on Earth that are not suitable for detecting neutrinos of the other family. Finally, neutrinos can provide important information about the most remote areas of the Universe since they interact little with the matter they encounter on their way. So from this point of view, they are excellent messengers, provided that they can be detected in sufficient numbers to reconstruct the information they carry.

Returning to quantum field theory, one of the basic assumptions is that the fields associated with forces are quantized and particles are excitations of the field, photons are indeed excited states of the electromagnetic field. More generally, fermions and bosons are respectively excitations of fermionic and bosonic fields. The new paradigm introduced by quantum field theory no longer provides a clear distinction between material particles and immaterial fields but a single physical quantity, the field, present in every point of space-time, whose excitations are the particles. In practice, a quantum field is treated as if it were made up of infinite quantum harmonic oscillators whose excited states are indeed the particles. Moreover, the fundamental forces of nature are transmitted through the emission and absorption of particles that are called force mediators, also excitations of the fields. For example, the Coulomb force between two electric charges is transmitted through the exchange of photons which are the mediators of the force electromagnetic. If we consider for example two electrons, from one of the two a photon leaves that is absorbed by the other and it is precisely this exchange of particles that produces the repulsive interaction between the two particles (Fig. 2.7).

In cases where the mediators are massless, the range of action of the forces is infinite, in the case of massive mediators the range of action is finite and is inversely proportional to the mass of the mediator.

But how are these mediator particles created? Do they come out of nothing? Actually yes: based on the energy-time uncertainty principle, it is possible to have an energy fluctuation in a short time as long as their product is of the order of Planck's constant.

Therefore, the greater the energy fluctuation, the shorter must be the lifetime of the virtual particle. In the case where the particle is massive, an energy fluctuation equal to at least its mass for the speed of light squared ($E = mc^2$) is required. To this energy fluctuation corresponds a time fluctuation which is precisely the lifetime of the particle, which will be so much shorter the greater the mass of the mediator particle. So even assuming speeds close to that of light, inevitably after a certain distance the virtual particle vanishes and therefore ends the action of the force of which it is mediator. Having the two

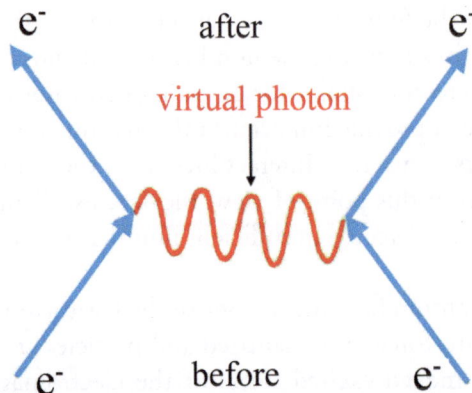

Fig. 2.7 Electromagnetic interaction. The repulsion force between the two electrons occurs through the exchange of a virtual photon. All interactions are characterized by the same mechanism

nuclear forces strong and weak a very small range of action, we deduce that the mediator particles are massive and in the case of the weak nuclear force they are even more massive compared to those of the strong nuclear force. In the case of the electromagnetic force, having the mediator (photon) zero mass, the range of action is infinite.

This new vision of fundamental interactions mediated by virtual particles stimulated some reflections and very suggestive hypotheses that led to a new concept of vacuum (*the quantum vacuum*) no longer considered as a portion of space in which neither matter nor radiation is present, but as something extremely dynamic in which, by virtue of the uncertainty principle, particles and virtual fields appear and disappear for a very short interval of time. After all, if we imagine a vacuum characterized by the absence of anything, we are implicitly stating that the energy of the vacuum is exactly zero but this is in disagreement with the energy-time uncertainty principle that imposes limits on the indetermination of energy, it cannot assume an exact value unless we assume an infinite indetermination over time.

A direct consequence of the quantum vacuum is a particular effect hypothesized by the Dutch physicist Hendrik Casimir and known precisely as the Casimir effect, a spectacular macroscopic evidence of the quantum vacuum. This effect predicts that two metal plates, if placed at a very small distance from each other, attract each other but not due to gravitational or electrostatic effect but due to the difference in pressure exerted by the virtual photons that exist between the interior of the plates and the outside. In fact, since only photons with particular wavelengths related to the distances between the plates can exist inside the plates, these are fewer than the photons that are typically created outside; therefore, a pressure difference is established that tends to bring the plates closer. The dynamic Casimir effect instead predicts the conversion of virtual photons into real photons, due to a disturbance of the quantum vacuum like that generated by two mirrors approaching at speeds close to that of light. Essentially, this effect predicts the creation of light from nothing or, better yet, from the quantum vacuum. These extraordinary effects have been experimentally verified confirming the existence of the quantum vacuum. In 1997 at the laboratories of the University of California, the static Casimir effect was verified by measuring the attraction force between a sphere and a thin plate placed at a distance between 0.6 μm and 6 μm. In 2011, the dynamic Casimir effect was also verified, observing the creation of real photons. In this case, since a experimental system in which two mirrors moved at relativistic speeds could not be realized, a particular superconducting circuit was used in which a suitable transmission line of variable electrical length created the experimental conditions to observe the dynamic Casimir effect. In this experiment, it was possible to vary the electrical length of the transmission line at a speed equal to a substantial fraction of the speed of light, obtaining a system similar to the two mirrors in relativistic motion.

The concept of quantum vacuum could be at the basis of cosmological issues of extreme importance such as the accelerated expansion of the Universe beyond the expected. In other terms, it could justify the existence of a repulsive force similar to that imagined by Einstein and formalized with the introduction of the famous cosmological constant, considered by him a serious mistake (see Chap. 1). Moreover, just as light can arise from a fluctuation of the quantum vacuum, the Universe could have been born from a fluctuation of the quantum vacuum as theorized in the 70s and 80s by physicists Edward Tryon and Alexander Vilenkin, based on a hypothesis by Pascual Jourdan according to which a macroscopic object even of great size like a star could arise from a statistical fluctuation of the vacuum. Therefore, even in the world of the infinitely large, the uncertainty principle and the quantum vacuum are at the basis of very suggestive and disruptive hypotheses and theories. In this regard, before closing this interesting and necessary parenthesis on the quantum vacuum, it is worth remembering the theory of the famous English astrophysicist Stephen Hawking regarding the radiation emitted by black holes. The great astrophysicist in 1974 hypothesized that black holes are not black at all in the sense that in addition to swallowing everything, including light, they also emit a radiation known as *Hawking radiation*. On the event horizon of a black hole (i.e., the surface beyond which nothing can escape from a black hole), according to Hawking's theory, pairs of particle-antiparticle would be created due to a quantum vacuum fluctuation, one of which would be swallowed by the black hole and the other would be emitted (Hawking radiation). Since energy must be conserved, the particle that falls into the black hole must have an energy equal and opposite to the one that escaped, which is equivalent to a negative energy that can be imagined as energy subtracted from the black hole. Therefore, even though the particle pairs were created by a quantum fluctuation, the enduring existence of the escaping particle implies an energy consumption by the black hole, which would inevitably consume/evaporate as it would convert all its mass into energy to emit Hawking radiation. Of course, we are talking about cosmological times, a black hole with a mass equal to that of the sun and completely isolated would take a time much longer than the age of the Universe (about 14 billion years) to consume itself. Moreover, it should be considered that the black hole, even if it emits radiation, at the same time continues to greedily feed by swallowing everything around it as it is not hardly completely isolated. Therefore, a complete evaporation of the black hole would only be possible for those of modest size, also because the power emitted by a black hole, due to its evaporation, is inversely proportional to its mass. The verification of the existence of Hawking radiation is not simple; however, some experiments have been carried out that seem to confirm Hawking's theory.

Now let's try to go through the fundamental steps that led to the standard model, or the quantum theory that explains the four fundamental interactions and the properties of the fundamental particles. Starting from the new paradigms of quantum field theory, in 1933 Fermi proposed the first theory of weak nuclear forces that predicted the point or contact interaction of the three fermions involved (neutron, electron and antineutrino) in beta decay. According to the theory, the emission of an electron and an antineutrino by some atoms was due to the transformation of a neutron into a proton. This transformation took place inside the nucleus thanks to a new force later called weak nuclear force and it was in the nucleus that the electron was generated and escaped from the atom. In some way a process similar to the generation of photons when an electron moves from an orbital with higher energy to one with lower energy. Fermi's theory allowed to calculate with good accuracy the energy of the electrons emitted in beta decays and to predict that the new weak nuclear force was about 10,000 times less intense than the electromagnetic one. Fermi submitted the results of his theory to the prestigious journal "Nature", which did not accept the work arguing that the article was too speculative and of little interest to the journal, only to regret it later admitting the serious editorial mistake. The Italian scientist published the theory in the Italian journal "Il nuovo cimento" and in the German journal "Zeitschrift fur

Phisik.", but disappointed and upset by the rejection of the prestigious journal, he decided to devote more time to the experimental aspects of physics that soon led him to win the Nobel Prize in 1938 for his studies on nuclear reactions and the discovery of new artificial radioactive elements.

Enrico Fermi, leader of the famous boys of via Panisperna, pride of Italian physics, was a formidable physicist, one of the few to juggle skillfully between theoretical and experimental physics and capable of giving fundamental contributions in both fields. Probably the greatest Italian physicist since the time of Galileo.

Subsequently, Fermi's theory was modified by introducing new mediating particles that were exchanged during the beta decay process, so no longer a direct interaction between the particles but through virtual particles.

Inspired by Fermi's theory of weak interactions, the Japanese physicist Hideki Yukawa in 1934 proposed the first theory of the strong nuclear force that, in analogy with the electromagnetic one mediated by the photon, was transmitted through the exchange of a particle with a mass of about 140 MeV (about 280 times the mass of the electron), called *meson π or pion*. Therefore, in addition to Fermi's theory, Yukawa hypothesized the existence of exchange nuclear forces mediated by a virtual particle. He also introduced a potential, known as Yukawa's potential, which took into account the very small field of action of the strong nuclear force. In fact, unlike that of Coulomb for the electric force which is inversely proportional to the distance squared from the source, Yukawa's predicted a characteristic length that approximately coincides with the range of action of the strong nuclear interaction and a behavior with an exponential decrease, therefore a force that quickly goes to zero as soon as it exceeds its range of action. The particle hypothesized by Yukawa, initially confused with the muon (a heavy electron), was discovered in 1947 by Cesar Lattes, Cecil Frank Powell and Giuseppe Occhialini, using detectors based on nuclear emulsions in which, following the passage of a charged particle, silver bromide crystals ionize giving rise to the formation of dark grains with a diameter of about 0.5 mm. It is therefore possible to observe the trajectory of the particle and obtain fundamental information for identification and particle type. The three experimental physicists took the detector to the mountains and exposed it to cosmic rays; after a careful analysis of the trajectories left by the particles contained in the cosmic rays, they identified a particle whose characteristics matched those hypothesized by Yukawa. The Japanese scientist was awarded the Nobel Prize in 1949 for his theory of strong nuclear interaction. However, as we will see later, this theory will then be incorporated and replaced by more general theories in which the pion, as well as the proton and the neutron are composed of even more elementary particles.

Quantum electrodynamics (QED), developed by Richard Feynman, Julian Schwinger, Freeman Dyson, Sin-Itiro Tomonaga and Hans Bethe at the end of the 40s of the last century, is the quantum theory of the electromagnetic field and explains with unprecedented precision the discrepancies between the experimental measurements of the magnetic moment of the electron and the Lamb shift compared to the predictions of the theory of Dirac. The central aspect of the theory lies in the interaction of the electron with the quantized electromagnetic field and in the resolution of a big problem related to the infinite values that came out when the calculations were made. The problem of the infinities was solved by Bethe using a particular mathematical procedure called renormalization which involves the elimination of infinities through a redefinition of some constants. Despite this somewhat spurious way of proceeding, which is certainly not appreciated by purists of mathematical elegance in physics used to dealing with Einstein's field equation or Dirac's equation, the predictive capacity of QED is unprecedented and the technique of renormalization became a test to understand if a quantum field theory is valid or not. For the development of quantum electrodynamics, Feynman, Schwinger, and Tomonaga received the Nobel Prize in 1965. Particularly important was Feynman's approach which introduced some particular graphs known as *Feynman diagrams*; they represent the spatial and temporal evolution of the interaction between the particles and also simplify long calculations. These diagrams became a fundamental tool and widely used also in the other quantum field theories.

By now the path had been started and soon general theories would also be formulated for the strong and weak nuclear force although in this case it was fundamental the combined use of other theories of great beauty and importance known as *gauge theories* based on concepts of symmetry. The mathematical tool for studying symmetries is the *group theory* which soon will add to the already rich mathematical park of quantum physics.

As seen in the first chapter, Einstein was the first to introduce a principle of symmetry in modern physics; in fact, the first principle of special Relativity tells us that the laws of physics must be invariant with respect to Lorentz transformations. Subsequently, two other great scientists made a fundamental contribution to the use of symmetries in physics. The German mathematician Emmy Noether, perhaps the most important women in the history of mathematics, in 1918 formulated an important theorem known as Noether's theorem, which demonstrated that a symmetry of a physical system corresponds to a conservation law. Therefore, it is the symmetries that determine the conservation laws by constraining them. The great theoretical physicist Eugene Wigner argued that: "symmetries are laws that the laws of nature must respect".

Based on Noether's theorem, it can be demonstrated that invariance for translations, a consequence of the uniformity of space, gives rise to the conservation law of momentum, or that invariance for rotation due to the isotropy of space follows the law of conservation of angular momentum, or even invariance for temporal translations due to the uniformity of time follows the conservation of energy. In other words, the most important conservation laws arise from the fact that an experiment provides the same result if it is performed anywhere in space and at a distance of days, years or centuries.

But the German mathematician and physicist Hermann Weyl, perhaps considered the greatest aesthete of theoretical physics, goes further and at the end of the Twenties of the last century proposed to impose particular symmetries to derive the fundamental laws of physics (*gauge principle*), and in particular to unify the theory of general Relativity and that of electromagnetism. The bold venture failed but with the gauge principle Weyl had introduced the basic concept of the aforementioned gauge theories based on symmetries with respect to transformations in "internal" spaces.

In fact, it is required that the mathematical function that describes energy (*Lagrangian*) does not change shape with respect to transformations in these internal spaces and this necessarily implies the addition of some terms that describe the fundamental interaction in question including the type and number of mediators of the interaction. For example, if applied to electromagnetic force, the requirement of symmetry implies the existence of a force mediator which in this case is the photon.

Although Weyl had not succeeded in unifying General Relativity and electromagnetism, he was still proud of his theory and a few years later he said to Bethe, the physicist who had used the mathematical forcing of "renormalization" to eliminate the infinities from quantum electrodynamics: "In my research I have always strived to unite the true with the beautiful, but when I had to choose between the one and the other, I usually chose the beautiful".

But what are "internal" spaces? Let's take the simplest example: a proton and a neutron from the point of view of the strong nuclear interaction are identical; therefore, one can imagine an internal space with two dimensions (*strong isospin*) in which proton and neutron are the same particle but depending on how they are arranged in this space (up or down) they assume the character of proton or neutron. In other words, from the point of view of the strong nuclear interaction proton and neutron are two sides of the same coin, depending on how the coin is oriented we observe the proton or the neutron. In this sense we can say that the strong nuclear force is invariant for rotations in the space of strong isospin. In fact, if in this space we rotate the proton by

180° we obtain the neutron which from the point of view of the strong nuclear interaction is identical to the proton.

In the 1950s, the two theoretical physicists Chen Ning Yang and Robert Mills formulated a gauge theory to explain the strong interactions starting precisely from a symmetry for rotation in the two-dimensional internal space of strong isospin. The theory was not successful because the imposition of gauge invariance led to the existence of three massless force mediators, but having the strong interaction a very small range of action the mediating particles could not be massless.

At the beginning of the 60s, the American physicist Sheldon Glasgow revisited the theory of Yang and Mills, but this time to explain the weak force for which there were strong indications that the number of mediating particles were three. Once again the problem of the mass not predicted by the theory extinguished interest in Glasgow's theory.

The turning point came in 1964 when the British physicist Peter Higgs and independently Francois Englert and Robert Brout, introduced a mechanism that allowed particles to acquire mass. The mechanism known as the Higgs mechanism was based on the existence of an omnipresent field whose mediator was the famous Higgs boson having a mass 125 times greater than that of the proton and discovered in 2012 at the CERN laboratories about 50 years after its prediction. So as in the case of the electromagnetic field, the photon is the mediator of the forces and can be considered as an excited state of the electromagnetic field, in the same way the Higgs boson is an excited state of the Higgs field and its discovery has confirmed the mechanism that underlies the acquisition of mass.

When trying to explain the Higgs mechanism without using the due physical-mathematical formalism, it is very likely that one provides a misleading and mystified picture. Nevertheless, like other sources of dissemination on the topic, we will try to give a very intuitive explanation using some metaphors with the macroscopic world.

We could imagine the Higgs field as a calm sea: if we consider a catamaran, there is less water resistance while a merchant or cruise ship certainly offers greater resistance. The catamaran and the cruise ship in our metaphor represent the particles. These must necessarily cross the Higgs field each with a different resistance and it is precisely this resistance that identifies the mass of the particles. Particles with small or zero masses like the photon and the electron (catamaran) are little interacting with the Higgs field while the heavier ones like the proton and neutron (cruise ship) interact more and therefore the mass is greater.

We can also imagine a tub full of oil (Higgs field) in which we immerse small spheres of negligible mass and covered with a material with a different degree of oil impermeability. The spheres that absorb more oil will have a greater mass while those that are less absorbent will have a lesser mass or in case of perfect oil repellency, they will have a negligible mass. In this analogy the spheres more soaked in oil correspond to particles with greater mass while those oil-repellent to particles with lesser mass.

We can therefore affirm that the particles that interact a lot with the Higgs field are the most massive, while the light particles like the electron interact little and those without mass like the photon do not interact at all. Surely the reader will not miss a certain analogy between the Higgs field and the ether, both pervade the entire space and both have been introduced to make sense within the theory to physical quantities (mass and electromagnetic waves). However, in the case of the Higgs field there is unquestionable experimental evidence while in the second case there has never been any proof of its existence indeed there has been a theory (special Relativity) that has justified its non-existence in an excellent way.

Let's now try to give a slightly more formal explanation using a more general phenomenon known as *spontaneous symmetry breaking*, introduced in 1961 by Yoichiro Nambu, Giovanni Jona-Lasinio and Jeffrey Goldstone for elementary particles inspired by an analogy with superconductors, materials that we will deal with in the next chapter.

What is meant by symmetry breaking? To understand this, let's first consider a simple example. A perfect homogeneous and monochrome circular disc is invariant for rotation in the sense that if we rotate it by any angle, it always looks the same. If we cut it in half, it will no longer be invariant for any rotation but only for 360-degree rotations. We have therefore broken the symmetry. Another example is that of a rubber cylinder. The cylinder is clearly symmetric for any rotation around its own axis. If we apply a weak force on the upper surface of the cylinder, the symmetry is preserved as it does not deform. But if the intensity of the force exceeds a critical value, the cylinder will start to bend on itself assuming a curved shape. In this case the symmetry for rotation is lost and we are therefore in the presence of a symmetry break.

In the case of the Higgs field something similar happens. The mathematical function that describes the potential associated with the Higgs field has the shape of a sombrero (Fig. 2.8) that enjoys a perfect symmetry for rotations around its own vertical axis. But the most important peculiarity of the Higgs potential is that it has the minimum energy at a field different from zero; therefore, the states of minimum energy are characterized by the presence of the Higgs field and at the relative maximum point of the potential (the top of the sombrero), the field is zero.

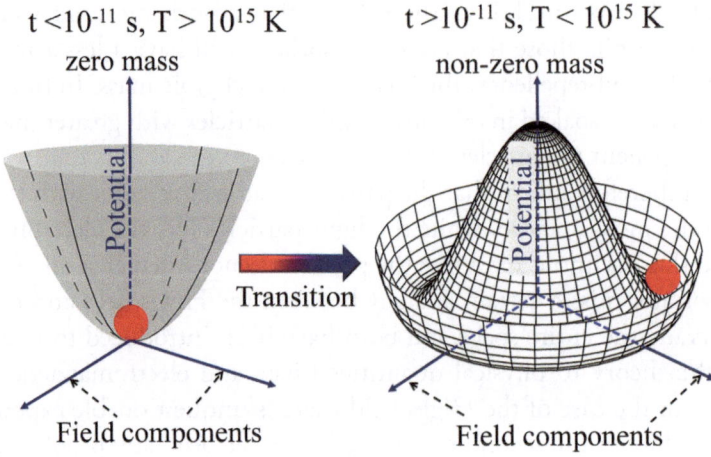

Fig. 2.8 Representation of the potential associated with the Higgs field. In the early moments of the Universe, the Higgs field was essentially zero as the minimum energy corresponded to a null field value. Following the transition from the paraboloid potential (left) to the sombrero potential (right), which occurred after about one hundred billionth of a second (10^{-11} s) from the big bang, the presence of the infinite minimum energy at a field different from zero implied a spontaneous symmetry breaking that allowed matter to acquire mass

If instead we consider a potential paraboloid or bowl-shaped (Fig. 2.8), the minimum of the potential corresponds to a null value of the field.

An example is given by the electromagnetic field in which the states with zero energy correspond to the total absence of electric and magnetic fields. Even though the two potentials represented in Fig. 2.8 are both symmetric for rotation around the vertical axes, in the case of the sombrero potential there is a spontaneous symmetry breaking. In fact, the only point of minimum energy of the paraboloid potential becomes, in the case of the sombrero potential, the entire groove containing infinite points of minimum energy. If we imagine a particle in the respective bottoms of the potentials, in one case it can only oscillate around the only point of minimum energy (bowl-shaped potential), in the other case in addition to oscillating along the side walls it can also slide along the groove (sombrero potential). Moreover, the choice of a particular point of the groove where the particle is positioned is evidently a symmetry break. It is as if we put a label inside the groove of the hat with the consequent loss of symmetry with respect to any rotation around the axis. In fact, if the hat is rotated, the position of the label changes and consequently we deduce that a rotation has been made. Conversely, in the absence of a label we have no way of realizing that the hat has been rotated.

In the early moments of the Universe, due to the high temperature, the aforementioned potential had a shape similar to a paraboloid, which is why in the states of minimum energy corresponding to the minimum of the curve, the Higgs field was essentially zero.

When the temperature dropped below the critical value of about 10^{15} °C (10^{-11} s after the Big Bang), there was a transition from the paraboloid potential to the sombrero one, whose maximum is a point of instability. In fact, a minimal quantum fluctuation causes the system to decay into one of the infinite points of the groove with the lowest energy. In other words, a particle placed at the maximum immediately rolls into the groove of the sombrero. On the other hand, in nature all systems tend to the minimum energy.

The consequent spontaneous symmetry breaking led to the appearance of two types of particles: one without mass characterized by an oscillation along the groove of the sombrero potential and the other having mass whose oscillation occurs along the lateral walls of the groove. The reason why the particle's mass is born in the groove of the sombrero potential can only be understood by addressing the issue with the appropriate formal tools of theoretical physics but this explanation, although incomplete, is the closest to the theory of Higgs, Englert and Brout.

To complete this more formal parenthesis, it should be said that the mechanism by which a spontaneous symmetry breaking was necessarily accompanied by the appearance of particles (*Goldstone bosons*), was a theoretical result already consolidated (Goldstone theorem) but according to the theory the particles were massless. The intuition and merit of Higgs, Englert and Brout was to have extended the spontaneous symmetry breaking to a *local gauge theory*, in which the massless Goldstone bosons combine with the also massless gauge mediator bosons originating particles having mass.

We can therefore affirm that the symmetries of nature play an important role but equally fundamental are the symmetry breaks thanks to which the Universe we know with bodies endowed with mass and without antimatter was born. In this regard, the physicist B.G. Wybourne, an expert in symmetries and group theories, argued: "What an imperfect world it would be if every symmetry were perfect"

Following the fundamental discovery of the Higgs boson that is a signature of the mechanism at the base of the origin of mass, Peter Higgs and Francois Englert won the Nobel Prize in 2013 for having predicted it about 50 years earlier, Robert Brout died in 2011 and unfortunately he had neither the satisfaction of seeing his theory verified, nor that of winning the Nobel Prize which by regulation cannot be awarded posthumously.

We close this necessary section dedicated to the Higgs boson with a curiosity. Often the Higgs boson is called the God particle, alluding to creation, since this particle is linked to the origin of mass. In reality, this nickname comes from the unwanted title of a very successful popular science book written in 1993 by the Nobel laureate for physics in 1988, Leon Lederman, *The God particle*. The original title proposed by the author to the publisher was *The goddamn particle* for the great difficulty in being discovered, but for marketing reasons the publisher convinced Lederman to use a more attractive title and the adjective *goddamn* became *god* hence the final title, the God particle.

Having solved the problem of the mass of the force mediators, interest in gauge theories was rekindled and in 1967 Steven Weinberg, Abdus Salam and Sheldon Glasgow formulated the definitive theory of the weak interactions including also electromagnetic interactions in a unified theory known as electroweak theory in which, in addition to the photon, mediator of the electromagnetic force, three massive particles W^+, W^- and Z^0 appeared mediating the weak nuclear force.

Just as electricity and magnetism are two sides of the same coin, according to the electroweak theory at very high energies corresponding to a temperature of 10^{15} °C, all electromagnetic phenomena including light and much of radioactivity, are manifestations of a single force namely the electroweak force. For the aforementioned theory, the three scientists were awarded the Nobel Prize in 1979. The three particles mediating the weak nuclear interaction were discovered in 1983 at the CERN laboratories under the direction of the Italian physicist Carlo Rubbia who for this discovery received the Nobel Prize in 1984.

Apart from the gravitational force for which there is still no consolidated quantum theory, there remains the strong nuclear force that requires some considerations.

Starting from the discoveries of the positron and the neutron in 1932, new particles were discovered at a surprising rate. The number of particles was such that it was even difficult to remember their names. Enrico Fermi to a student's question about the names of the particles replied: "boy, if I could remember the name of all these particles I would be a botanist".

In 1960, another giant of modern physics, the American physicist Murray Gell Mann put some order by doing something similar to what about a hundred years earlier the Russian chemist Dmitri Mendeleev had done for chemical elements. Gell Mann, based on symmetry considerations, organized subatomic particles into groups of 8 or 10 and to which he gave the name of *eightfold way* inspired by the noble eightfold path of Buddhism. In 1964, *annus mirabilis* of elementary particle physics, Gell-Mann and independently

George Zweig introduced the quark model, a term of pure fantasy inspired by a nonsense phrase contained in the novel *Finnegans Wake* by James Joyce (*Three quarks for Muster Mark!*). The model predicted that all particles subject to the strong nuclear force (hadrons) were formed by two or three truly elementary particles, *the quarks*. For example, the proton and the neutron are made up of three quarks (Fig. 2.9).

In the same year, Oscar Greenberg, observing that two identical quarks could not occupy, due to the Pauli exclusion principle, the same state as happened in protons and neutrons, introduced a new quantum number that he named *color*. The color, which has nothing to do with the colors we see with our eyes, could assume three values, identified with red, blue and green, in line with the fantasy name. A proton or a neutron is composed of three quarks with different colors in order to satisfy the Pauli principle.

From the point of view of the strong nuclear force the quarks are perfectly equivalent. Moreover, the quarks were the first particles to have a fractional charge, in particular the quark down has a charge of $-1/3$ e (the electron's charge e is equal to 1.6×10^{-19} Coulombs) while the quark up has a charge equal to $2/3$ e. In the case of the proton, there are two quarks up and one quark down so as to have a total charge equal to e, while in the case of the neutron we have one quark up and two quarks down with a total charge of zero (Fig. 2.9).

The existence of quarks was first demonstrated in 1968 at the Stanford linear accelerator in California (USA), through an experiment similar to that of Rutherford in which instead of using alpha particles to bombard a gold foil, they collided a high-energy electron with a proton. For these extraordinary successes in understanding the most hidden structure of matter, Gell Mann

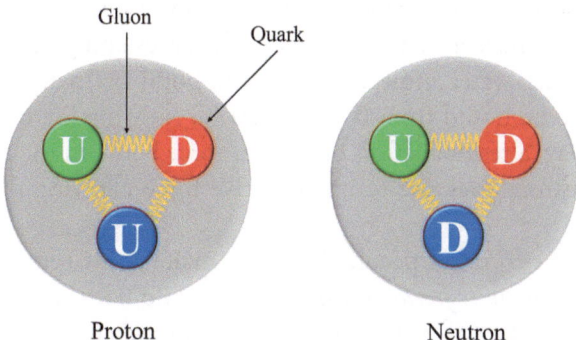

Fig. 2.9 Structure of the proton and neutron: both are made up of three different colored quarks. The quarks exchange gluons which are the mediators of the strong nuclear force

received the Nobel Prize in 1969 and the three experimental physicists Jerome Friedman, Henry Kendall and Richard Taylor who demonstrated the existence of quarks won the Nobel Prize in 1990.

At this point, the time was ripe to develop a theory for the strong nuclear interactions and as for the weak interactions the most natural choice was to use a gauge theory in which the internal space was the three-dimensional one of the three colors. In this space, the three quarks are completely equivalent and depending on the position they assume one color rather than another. By imposing the invariance of the function mathematical that takes into account the energy (*lagrangian*) with respect to this symmetry, they found 8 mediators of the nuclear force to which was given the name of *gluons* (from *glue*), the whose mass is zero. The quarks inside a nucleon exchange continuously the gluons which, being also colored, make change the color of the starting and arriving quarks. The theory known by the name of *quantum chromodynamics* was developed between the end of the years '60 and the beginning of the years '70.

At this point it is spontaneous to observe that the nuclear interaction strong, having mediators with zero mass, should have a radius of infinite action; instead, we said that the radius of action of the nuclear force is very small. How come? Are they different forces those that hold together the quarks compared to the attractive ones between two protons, two neutrons or a proton and a neutron? Actually, it is the same force, but the one that holds together the nucleus of an atom is a residual force called indeed residual nuclear force. From the point of view of the strong interaction, a proton or a neutron are neutral objects: in fact the three quarks that compose them have different colors (blue, red and green) and together they give a neutral color just like it happens for an atom or a molecule in which the number of positive charges and those negative are equal. However, even if neutral, a molecule like that of water can have a small imbalance of charges and give rise to chemical bonds with other water molecules, it is the bond or hydrogen bridge that allows water to be liquid from 0 °C to 100 °C allowing life on planet Earth. In the same way, a residual effect of nuclear force between the quarks inside a proton or neutron gives rise to an attractive force allowing the existence of atomic nuclei.

Quarks have two other peculiarities that make them very particular: *confinement* and *asymptotic freedom*. Isolated quarks do not exist, they have never been observed; in fact, if you tried to separate a single quark inside a neutron or a proton, the force to be applied would increase vertiginously as the quark is moved away from the other two quarks, a sort of force elastic that increases as the elongation of the spring increases.

Quarks are therefore destined to remain confined in their shell together with other quarks, continuously exchanging gluons. From an energy point of view, it is more convenient to form a pair of quark and antiquark rather than tearing a quark. So such an attempt would culminate in the creation of a pair of particles obtained from the conversion of the energy provided to tear away the quark. A bit like what happens with a magnet: it is not possible to isolate the north pole from the south, if a magnet is broken in two, two magnets are obtained.

From a theoretical point of view, the confinement of quarks has not yet been solved and is believed to be closely linked to a problem of physics-mathematics (*Yang Mills theory with mass gap*) part of the seven millennium problems identified by the Clay Mathematics Institute which has offered a prize of one million dollars for anyone who can solve one of the seven problems of the millennium.

If instead the quarks approach each other, their interaction energy decreases more and more to the point that very close quarks to each other are practically free, hence the term *asymptotic freedom*. The phenomenon is apparently in contrast to the behavior of other forces that decrease as the distance increases and increase as it decreases.

However, in 1972, David Gross, his student Frank Wilczek and independently David Politzer explained the phenomenon within the framework of gauge theories. The discovery of the three physicists, awarded with the Nobel Prize in 2004, proved to be fundamental for the theory of strong nuclear interactions, definitively demonstrating the effectiveness of quantum field theory, questioned by the infinite values present at very small distances like those that appeared in quantum electrodynamics.

The last aspect of quarks that is certainly worth mentioning is their mass. As mentioned, a nucleon (proton and a neutron) whose mass is about 1 GeV/c^2 is made up of three quarks whose sum of the masses is incredibly smaller than the mass of the nucleon. In fact, the mass of an up quark is 2.3 MeV/c^2 and that of a down quark is 4.8 MeV/c^2; in the case of the proton formed by two up quarks and one down, the sum of the quark masses is 9.4 MeV/c^2 or about one hundredth of the mass of the proton. The same applies to the neutron. Since electrons, having a mass equal to about 0.5 MeV/c^2, contribute negligibly to the total mass of atoms and therefore of matter, it is legitimate to ask what the other 99% of the mass that does not return in the accounts is made of. The answer comes again from Einstein's famous formula $E = mc^2$, it is binding energy that converts into mass. We can therefore affirm that we are essentially made of condensed energy and that the Higgs mechanism accounts for 1% of the stable matter.

QED, the electroweak theory and quantum chromodynamics constitute the backbone of the standard model that for over 50 years best describes elementary particles and fundamental interactions.

The table shown in Fig. 2.10, summarizes the results: there are three families of particles of which only one constitutes the stable matter (first column), consisting of quarks up, quarks down and electrons; the particles of the other two families are unstable and decay almost instantly into the particles of the first family. For example, a muon, belonging to the second family of leptons, decays into an electron, an electron antineutrino, and a muon neutrino in about two millionths of a second.

These unstable particles are created and observed only in large particle accelerators such as CERN in Geneva, Fermilab in Chicago or SuperKEKB in Tsukuba, Japan or to a lesser extent in cosmic rays.

The fundamental forces are four of which two unified in the electroweak interaction; the force mediators are eight (gluons) for the strong nuclear force, three for the weak nuclear force (intermediate bosons, W^+, W^- and Z^0) and one for the electromagnetic force (photon). Finally, there is the Higgs boson which allows particles to acquire mass. Therefore, the truly fundamental particles (at the moment) are 25 of which 12 are quarks and leptons, the photon,

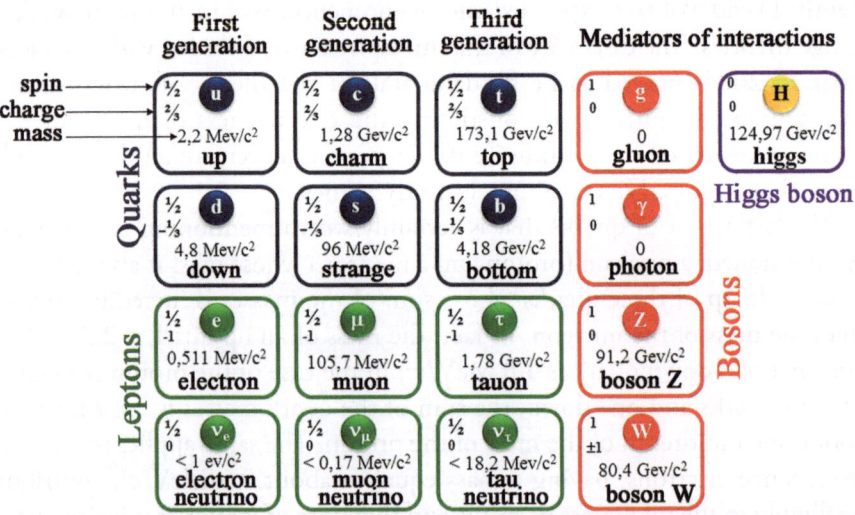

Fig. 2.10 Summary table of the standard model. The matter surrounding us is made up of particles from the first family (up and down quarks, electrons). The mediators of the four forces are massless in the case of the strong nuclear force (gluons) and the electromagnetic force (photon), the three massive W^+, W^- and Z^0 bosons are instead responsible for the weak nuclear force. Finally, there is the Higgs boson, the quantum of the Higgs field, which allows particles to acquire mass

8 gluons, 3 intermediate bosons and the Higgs boson. Naturally to these must be added the corresponding antiparticles which however, as said, do not constitute ordinary matter.

The force of gravity remains outside the standard model for which there is not yet a consolidated quantum theory and which together with other unresolved mysteries such as dark mass and energy or the initial asymmetry between matter and antimatter represents the challenge of fundamental physics in the coming years.

Of course, theoretical physics in the last half century has not stopped at the standard model but has produced fascinating theories such as the grand unification theory or theory of everything, supersymmetry, the string theory and loop quantum gravity to explain the limits of the standard model and gravitational interaction from a quantum point of view. However, unlike the consolidated standard model whose predictions are continuously verified with unprecedented precision, the aforementioned theories do not yet enjoy satisfactory verifications because in many cases it is difficult to realize the experimental conditions to verify them. Therefore, while solving some problems, they cannot be considered on the same level as the standard model which for over 50 years remains the most reliable explanation of elementary particles and fundamental interactions.

3

Condensed Matter and the Technological Impact of the First Quantum Revolution

In this chapter we will show the potential of quantum mechanics also for the study of more complex systems than atoms and elementary particles, like condensed matter formed by billions of billions of atoms. In this context we will have the opportunity to talk about very particular materials in which quantum aspects manifest themselves even at a macroscopic level. The chapter concludes with the most important applications of quantum physics, which have led to a real technological revolution that has changed our way of life.

3.1   Quantum Mechanics and Condensed Matter

All the matter that surrounds us, including ourselves, is made of atoms; therefore, it is legitimate to ask whether quantum mechanics can also be applied to systems composed of more atoms, such as molecules, solids and even more complex systems. Even though modern supercomputers and future quantum computers can help simulate with the laws of quantum mechanics fairly complex systems like molecules of biological and pharmacological interest, at the moment it is excluded that one can write and solve Schrödinger's equation or that of Dirac for a human being, or for one of his organs. These are extremely complex systems to be studied with the methodologies of quantum mechanics. Moreover, as we will see in the next chapter, in the case of very complex macroscopic systems, quantum effects are suppressed by particular phenomena of interaction with the surrounding environment, known as *quantum decoherence*. On the other hand, just as Relativity reduces to classical mechanics for speeds negligible compared to that of light, quantum mechanics also

© The Author(s), under exclusive license to Springer Nature Switzerland AG 2025
C. Granata, *A Journey into Modern Physics*, https://doi.org/10.1007/978-3-031-77775-2_3

reduces to classical when the quantities involved are much larger than Planck's constant. In more mathematical terms, we can say that classical physics is the limit for c (speed of light) tending to infinity and h (Planck's constant) tending to zero; in fact, if h were zero we would not have a quantum world, just as if c were infinite we would not have Special and General Relativity.

However, excluding particularly complex systems, quantum mechanics has been used with great success to explain the behavior of molecules, solids, fluids, gases and phase transitions, that is, that field of physics known as condensed matter physics. And it is precisely from this field, perhaps less fascinating compared to the physics of elementary particles, that the applications of greatest technological impact of quantum mechanics have been born.

Let's clarify that typically the quantum study of aggregated matter does not deal with macroscopic objects but is focused on its elementary constituents (electrons and nuclei), which determine the macroscopic properties. However, as we will see later, there are some materials whose extraordinary properties require a macroscopic quantum interpretation.

Moreover, since the speeds of electrons are much smaller than the speed of light, it is legitimate to use non-relativistic quantum mechanics. For example, in a hydrogen atom the speed of the electron is less than one hundredth of the speed of light.

With the birth of quantum mechanics, the interests of most physicists in the early 1900s were directed towards the study of atoms and their constituents while the study of solid state physics, which was later incorporated into the broader field of condensed matter physics, was essentially linked to practical applications especially those related to the new technologies that were emerging from the now consolidated field of electromagnetism.

In this context, it is not difficult to imagine that between the end of the nineteenth century and the beginning of the twentieth century, electrical phenomena played a very important role and there was an interest in investigating the mechanisms related to electrical conduction in metals. In 1827 the German physicist and mathematician Georg Ohm formulated empirical laws that bear his name. The first and most famous Ohm's law tells us that the electrical voltage at the ends of a metal is proportional to the current that flowing to it by a constant that is called *resistance* and represents a real resistance to the passage of electric current. Therefore, for a given voltage applied to a metal, the greater the resistance, the smaller the current. Because of this sort of electrical friction, the passage of current in a material produces heat, dissipating energy in the form of thermal energy (*Joule effect*). The power dissipated is proportional to the resistance and the square of the current flowing in the material. Ohm's second law tells us that resistance is proportional to the length of the metal and inversely proportional to its cross-section by a constant (*resistivity*) that depends on the metal. Like friction, which is essential for many things including walking or driving, electrical resistance is not always something undesirable, in fact, it is the basis of many applications; just think of various appliances such as the oven, hairdryer, electric heater, and washing machine that use appropriate resistors to produce heat when crossed by electric current.

The first microscopic theory of electricity conduction in a metal was developed in 1900 by the German physicist Paul Drude, who imagined the metal as a sea of electrons free to move on a substrate of ions made up of atomic nuclei and therefore much heavier. These were the outermost electrons, also called *valence* electrons, and therefore less bound to the nuclei. This gas of electrons continuously and randomly hits the ions; therefore, if an external force is applied through an electric voltage, the motion of the electrons is determined by the force of the electric field due to the applied voltage and by a friction force due to the continuous collisions which tends to stop the electrons. According to Drude's model, there is a sort of compensation between the two forces and therefore the resulting motion of the electrons is a zig-zag motion but at a constant speed, directly proportional to the applied voltage. Since speed is also proportional to the electric current, Drude deduced that voltage is proportional to current, i.e., Ohm's first law. However, Drude's model failed to explain another important characteristic of metals, thermal conductivity, or the ability of metals to transport heat, which, as we know from everyday experience, is highly efficient. In particular, the specific heat per mole was greater than the experimental one.

The first quantum theory of electrical conduction was developed in 1928 by the German physicist Arnold Sommerfeld, who started from Drude's model of free electrons but took into account a fundamental quantum aspect: Pauli's exclusion principle and the consequent Fermi-Dirac quantum statistics to which electrons are subject as fermions. Sommerfeld's theory was able to explain Ohm's law, thermal conductivity and its dependence on temperature, as well as some experimental effects observed in metals such as the Seebeck effect, i.e., the existence of an electric voltage at the ends of a metal produced by a temperature difference in the metal. Furthermore, the empirical law of Wiedemann-Franz was also explained, according to which the ratio between thermal conductivity and electrical conductivity is directly proportional to temperature.

In short, it was enough to apply the newborn laws of quantum mechanics to an extremely simple model like Drude's to explain numerous experimental phenomena and empirical laws related to electrical and thermal conduction in metals.

This certainly ignited enthusiasm in theoretical physicists who hoped to explain the behavior of solids with the laws of quantum mechanics. However, there were many doubts about Sommerfeld's theory, both due to its limitations and to basic conceptual ones. It was rather difficult to accept a model of free electrons since inside the solid there are ions with which the electrons should continuously interact.

A fundamentally important step towards a quantum theory of solids was made at the end of the 1920s by a student of Heisenberg, the Swiss physicist Felix Bloch, who was assigned a doctoral thesis on metals. The young physicist, awarded the Nobel Prize in 1952, demonstrated that for an ordered structure of ions (a *lattice*), the solution to Schrödinger's equation is given by the product of a simple periodic function for a free wave. This theorem, in addition to drastically reducing skepticism towards Sommerfeld's theory, will form the basis for developing, in the following years, a theory of fundamental importance for the quantum approach to condensed matter: the *band theory*. As can be inferred from the name, in solids we move from energy levels, in the case of isolated atoms, to energy bands in the case of a very large ordered set of atoms. This means that in solids, electrons are not forced to assume a well-defined level of energy but can have any energy within the range that outlines the band. Intuitively, the birth of bands can be understood by imagining two very close identical atoms, in this case the orbitals related to the same energy levels can merge and the electrons will be delocalized in both orbitals; therefore, what was previously a single energy level of an atom occupied by a single electron becomes a pair of levels occupied by both electrons. However, this mechanism leads to a very slight thickening of the energy level, that is, from the fusion of the two identical orbitals two distinct energy levels are born, but so close in numerical terms that they can be considered overlapping. If we now imagine three identical atoms and repeat the reasoning above, from the same energy level of the single atom, we will have three very close energy levels and so on. If we finally consider a very large number of atoms like those present in a solid, let's say a number equal to Avogadro's number (6×10^{23} atoms), each energy level of the single atom transforms into a large number of almost overlapping levels that give rise to an energy band. Therefore, there will be as many bands as there were energy levels in the starting atoms. In analogy to valence electrons, the valence band is the one with the highest energy completely occupied by electrons, while the conduction band is the one with the lowest energy empty or partially occupied. The band theory allows for a new classification of solids into metals, insulators, and semiconductors (Fig. 3.1). In

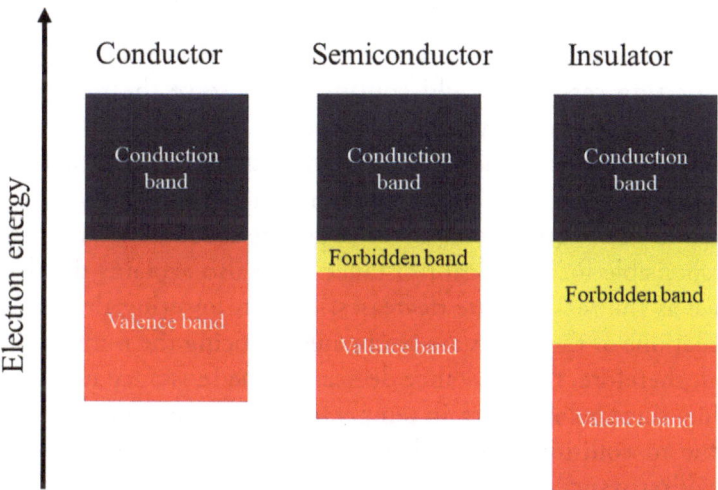

Fig. 3.1 Representation of conduction and valence bands in conductive, semiconductor and insulating materials. We move from conductors, where the bands overlap, to semiconductors, where there is a small forbidden band or gap (a few eV), to insulators with a substantial gap (greater than 5 eV)

particular, in the case where the valence band and the conduction band overlap, the electrons are free to move through the solid and therefore we have a metal; in the case where the two bands are separated by a forbidden band (*energy gap*), the electrons are blocked in the valence band and depending on the separation of the bands we refer to them as semiconductors (gap of the order of a few *eV*) or insulators (gaps greater than 5 *eV*). The band theory also resolved doubts related to Sommerfeld's free electrons model, as it incorporated into the model the interaction between electrons and lattice atoms.

However, quantum calculation did not predict the phenomenon of dispersion/scattering (better known as *scattering*) of electrons following collisions with the atoms of the lattice, a phenomenon at the basis of the Drude model to explain resistance in metals and, therefore, Ohm's law. In fact, in the presence of lattice imperfections, scattering processes occur producing a dissipative mechanism that underlies the resistivity of metals.

What are the imperfections of an atomic lattice? Certainly the impurities that, no matter how pure a metal may be, are always present, or the vacancies characterized by the absence of atoms, or also interstitial defects that occur when an atom is located in the space between two atoms that occupy the lattice, or finally lattice deformations.

If we had an ideal metal with a perfect and homogeneous lattice, would there be no electron scattering and therefore no electrical resistance? Even in this limiting case, we would continue to observe the resistance of the metal. And here another type of imperfection comes into play, impossible to eliminate: the vibrations of atoms and nuclei due to thermal agitation. These, too, are imperfections as the atoms and nuclei, oscillating, are not stationary in their equilibrium position giving rise to the scattering mechanism responsible for electrical resistance. This also explains the decrease in resistance as the temperature decreases. In fact, lowering the temperature, the oscillations of the atoms are attenuated reducing the number of scatterings and, therefore, the resistance decreases. The resistance never assumes a null value, since below a certain temperature, there remains a residual resistance due to collisions with other types of imperfections. The complete absence of resistance is instead typical of another phenomenon that we will talk about later.

Despite the successes of band theory, there was not much theoretical interest in the issues of condensed matter physics. Pauli, despite having dealt with it, giving suggestions and important contributions, argued that solid state physics did not belong to fundamental physics and that in some respects it was a "dirty" physics. Moreover, during the Second World War many physicists were involved in military and industrial research, further diminishing interest in the fundamental physics hidden behind solids.

In this context, however, a discovery was made that would trigger one of the most important technological revolutions of the last century: the invention of the transistor. In 1942 in the United States a project was funded on semiconductors aimed at applications on radar. The project led by Austrian physicist Karl Lerk-Horovitz, focused to study the properties of semiconductors and, particularly germanium, in order to develop an electronic device (the diode) made with semiconductors, with the aim of removing vacuum tubes and replacing them with smaller devices that consumed less energy. The excellent results of the project were a stimulus for the funding in 1945 of another project, at the Bell laboratories in New Jersey, under the direction of William Shockley and in which actively participated two other physicists John Bardeen and Walter Brattain. The three physicists focused on a particular three-terminal device made of germanium and in 1947 discovered that this device was able to amplify an electrical voltage. The device we are talking about is the transistor, which soon replaced all electronic devices made with vacuum tubes and started the phase of electronic miniaturization that continues even today. In 1956, the three American physicists were awarded the Nobel Prize for this important discovery.

3.2   Quantum Materials: The Extraordinary Manifestations of Quantum Physics at the Macroscopic Level

In the 1950s of the last century, the interest in the theoretical aspects of condensed matter was rekindled by the extraordinary behaviors of some forms of condensed matter, which had remained unexplained at least from the point of view of a consistent theory that took into account microscopic aspects and thus quantum ones. We refer to superconductivity, considered the most extraordinary manifestation of quantum mechanics at the macroscopic level.

Superconductivity, discovered in 1911 by the Dutch physicist Kamerlingh Onnes while measuring the resistivity of metals at very low temperatures, consists in the absence of electrical resistance in some materials when they are cooled to a temperature below a certain temperature called *critical temperature*. When we talk about the absence of resistance, we mean a sudden drop to zero of resistance, not a drastic reduction: in superconductors, the resistance is exactly zero at least within instrumental limits that are extremely low (less than one millionth of an ohm). Therefore, superconductors can also carry large amounts of current without any energy dissipation. The critical temperatures for superconducting metals are very low, less than −263 °C, therefore metallic superconductors must be cooled in liquid helium (−269 °C). This prevents them from being used as electrical cables to carry electricity in our homes, but as we will see in the next section, there are numerous applications of superconductors.

The total absence of electrical resistance is not the only property of superconductors: in 1933 Walther Meissner and Robert Ochsenfeld discovered that superconductors are *perfect diamagnets*, i.e., they completely expel the magnetic field from their interior (Meissner effect). This means that, if a magnet is placed on a superconducting material, the magnetic field does not enter the superconductor and, since magnetic forces are much stronger than gravitational ones, the magnet rises from the superconductor and begins to levitate, giving rise to the phenomenon known as *magnetic levitation*, which has nothing to do with the tricks performed by illusionists (Fig. 3.2). Considering these two exceptional properties, scientists immediately realized the potential applications of these materials and indeed the Nobel Prize for Onnes in 1913 did not take long to arrive. But equally extraordinary is the microscopic quantum theory that explained this particular phenomenon. Developed in the 1950s and published in its definitive form in 1957, the BCS theory, named after the three American physicists (John Bardeen, Leon Cooper, and Robert

Fig. 3.2 Photo of a small magnet levitating on a high-temperature superconducting disc cooled with liquid nitrogen (T = 77 K = −196 °C)

Schrieffer) who formulated it, predicts the formation of pairs of electrons, known as *Cooper pairs*. In fact, below the critical temperature, billions upon billions of electrons instantly pair up thanks to a weak attractive interaction, forming a set of Cooper pairs, the so-called superconducting condensate that flows in the superconductor without encountering any obstacles and, therefore, without any electrical resistance.

It is therefore a phase transition like those observed in everyday life, for example when water boils or ice liquefies in which we move from a liquid to a gaseous phase and from a solid to a liquid phase of water respectively. In the case of superconductivity, we go from a phase of almost free electrons to one of pairing.

The Cooper pair, being formed by two electrons with spin 1/2 and opposite directions, is actually a boson with spin equal to zero in the sum of 1/2 and −1/2 is zero, therefore the Pauli exclusion principle does not apply and all the Cooper pairs formed can be in the same quantum state, and can be described by a single macroscopic wave function. The size of these pairs is on the order of several tens or several hundreds of nanometers, so even 10,000 times larger than the radius of a hydrogen atom, this means that a long-distance bond is established between the two electrons of the Cooper pairs and that the size of a single pair can be greater than the distance between two pairs. Since the force between the pairs is very weak, if the thermal agitation exceeds a certain value, that is, if the temperature exceeds the critical one, the Cooper pairs break and superconductivity vanishes.

One might wonder: how do two electrons attract each other, perhaps even Coulomb's law no longer applies? No danger for Coulomb's law, two electric charges in a vacuum continue to repel or attract each other depending on the charge. The two electrons of the Cooper pair attract each other because the electrons interact with the nuclei of the material or with the quantized oscillations of the nuclei (*phonons*) and it is precisely from this interaction between electrons and phonons that the mysterious attraction between two electrons arises. In simpler and more intuitive terms, when an electron passes through a superconductor, it causes a slight deformation of the ionic lattice creating an area with a higher positive charge that attracts a subsequent electron, thus creating a pair of electrons mediated by the lattice. For the BCS theory, the three scientists won the 1972 Nobel Prize in Physics and Bardeen (the same as the transistor) was the only scientist in history to win two Nobel Prizes in the same discipline.

In 1986 Johannes Georg Bednorz and Alex Müller (winners of the Nobel Prize in Physics in 1987), discovered a new category of superconductors made with ceramic alloys consisting of a lanthanum, barium and copper oxide and characterized by a higher critical temperature (−238 °C). This discovery paved the way for the realization of so-called high critical temperature superconductors that allowed cooling through the use of liquid nitrogen (−196 °C), much more manageable and economical. In fact, the following year another ceramic material consisting of yttrium, barium, copper and oxygen (YBCO) with a critical temperature of −181 °C was discovered. Using these types of materials, a critical temperature of −140 °C was reached using a ceramic alloy composed of mercury, barium, calcium, copper and oxygen. Even higher critical temperatures have been obtained using materials subjected to very high pressures, in particular some hydrides subjected to pressures of over one million times atmospheric pressure show critical temperatures ranging from −70 °C to room temperature. The carbonaceous sulfur hydride, discovered in 2020, if subjected to the incredible pressure of 2.67 million atmospheres becomes superconducting with a critical temperature of 15 °C, practically room temperature. To get an idea of how great a pressure of several million atmospheres is, consider that the pressure one is subjected to in water at a depth of 1000 m is 100 atmospheres. Naturally, it is unthinkable, at the moment, to use such materials under such extreme conditions for practical purposes. Unlike superconducting metals, there is currently no exhaustive theory of the mechanism of high critical temperature superconductivity.

Another noteworthy phenomenon and in some respects similar to the previous one is the Bose-Einstein condensation named after the two physicists Satyendranath Bose and Albert Einstein who theorized it in 1924. Like superconductivity, Bose-Einstein condensation is a phenomenon linked to a phase transition: below a critical temperature, atoms with total integer spin (bosons), occupy the same quantum state at the lowest energy since the Pauli exclusion principle does not apply, and they assume a coherent and unitary behavior as if they were a single entity, showing quantum properties such as the wave nature of matter even at macroscopic level.

In order for this extraordinary phenomenon to be observed, the wavelength associated with the atoms must be comparable to their distance; since according to De Broglie's relation the wavelength of a particle is inversely proportional to its speed, it is necessary to lower the speed of the atoms, so that an overlap of the waves associated with the atoms can be obtained.

The first Bose-Einstein condensate was realized in 1995 by the two American physicists Eric Cornell and Carl Wieman and, independently by the German physicist Wolfgang Ketterle (awarded the Nobel Prize for Physics in 2011). The research group led by the two American physicists used a gas of rubidium atoms and cooled it to a temperature close to absolute zero $(0 K = -273.15 °C)$ using a particular technique based on laser cooling and magnetic trapping. By isotropically sending laser beams onto a gas of atoms, their slowing down is produced and therefore a decrease in temperature which, according to Boltzmann's statistical mechanics, is proportional to the speed with which the atoms move. So the more the motion of the atoms is slowed down, the more the temperature decreases and near absolute zero the atoms are almost stationary. The magnetic trap uses non-uniform magnetic fields to confine atoms that possess a magnetic moment like rubidium, which would otherwise tend to dissolve or simply fall due to gravity.

Below the critical temperature (about 170 billionths of K), the two scientists observed that the distributions of the atoms' speed were extremely peaked around zero, i.e., the atoms were virtually stationary unlike what happened for temperatures above the critical one where the distribution of speeds was isotropic and broadened (Fig. 3.3).

After creating two condensates with about five million sodium atoms each and placed at a distance of 40 millionths of a meter from each other, they made them interfere with each other observing a clear interference pattern typical of wave phenomena.

The Bose-Einstein condensate is considered the fifth state of matter that joins the solid, liquid, gaseous and plasmonic states, but unlike these latter it is the manifestation of a macroscopic quantum phenomenon that requires an interpretation exclusively within the framework of quantum physics. Unlike superconductivity, where the percentage of condensate is on the order of 10% of the electrons, in the case of the Bose-Einstein condensate a total

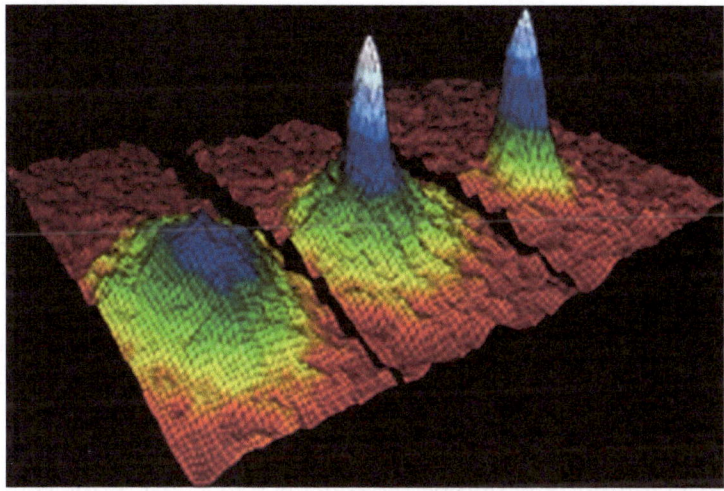

Fig. 3.3 Distribution of the speeds of cooled rubidium atoms for three temperatures close to absolute zero. Below a critical temperature, which in this case is 170 billionths of K, the distributions (center and right in the figure) are very peaked around zero indicating that most of the atoms are practically stationary

condensation of all the atoms contained in the cooled sample can be obtained, the size of which is a few tens of micrometers (hundredths of a millimeter).

This phenomenon has subsequently been observed in other atoms such as hydrogen, lithium, helium and potassium and on populations ranging from ten thousand to 100 million atoms. In addition to theoretical interest, the Bose-Einstein condensate could have interesting applications in quantum technologies or to create, thanks to their wave properties, extremely sensitive quantum atomic interferometers to be used for the detection of gravitational waves.

Another singular manifestation of quantum mechanics at the macroscopic level and in first approximation explainable within the framework of Bose-Einstein condensation is superfluidity, a phenomenon discovered in 1937 by Pëtr Kapica and, independently, by John F. Allen and Don Misener, stimulating the quantum study of fluids known as quantum hydrodynamics.

Superfluidity is the property of some fluids to flow in complete absence of viscosity. Just as a body in motion on a plane is subject to resistive forces due to friction with the plane and to air resistance, in the same way a fluid (liquid or gas) in motion is slowed down by viscosity, a kind of friction between the molecules of the fluid and, therefore, depends on the type of fluid considered. It is common experience that oil is more viscous than water which in turn is more viscous than air. Usually, for liquids, viscosity decreases with increasing temperature while for gases the opposite happens. In the case of some fluids such as helium, below a certain critical temperature (2.2 K), viscosity drops to zero as happens in superconductors for resistance. If observed at the moment of transition, liquid helium goes from a condition of continuous boiling to one of perfect calm with the sudden disappearance of all bubbles. The complete absence of viscosity implies that superfluid helium flows without any obstacle; if set in motion it could move without ever stopping (perpetual motion) and can pass through tiny holes that would block any liquid as shown in Fig. 3.4 (photo on the right). Moreover, it can form a very thin film (formed by a few rows of overlapping atoms) on the surface of objects and allow the passage of the superfluid through this film. By virtue of this effect, if you take a container wet with superfluid helium and partially immerse it in the superfluid, the latter climbs the external walls of the container and begins to fill it until the level inside and outside the container are equal. If the container is then lifted from the bath, the superfluid helium rises along the internal walls against the force of gravity and escapes completely emptying the container (Fig. 3.4, photo on the left).

Fig. 3.4 Superfluid liquid helium. The complete absence of viscosity allows the superfluid to escape from the container by climbing up the internal walls (photo on the left), or to pass through a ceramic bottom having ultra-fine porosity (photo on the right) that does not allow any fluid to pass through it (Credit: Condensed Matter Physics Center, UAM Madrid, Spain)

In addition to the absence of viscosity, superfluids are characterized by a very high thermal conductivity, that is, the ability to transmit heat and reach thermal equilibrium (same temperature) almost instantly. This means that, if we had an Olympic swimming pool filled with superfluid, an increase in temperature at one of the two edges of the pool, for example, due to an external heat source would be observed almost instantly also on the other edge of the pool 50 m away.

A direct consequence of this property is *the fountain effect*, considered the most spectacular effect of superfluidity. If a small capillary is inserted into the superfluid helium and it is heated even by only exposing it to light, the superfluid helium flows quickly into the capillary to balance the temperature. There is therefore a sudden rise of the superfluid helium up the capillary and the resulting fountain-like exit from it.

Even though the theoretical explanation of superfluidity has been the subject of a succession of fairly complex theories and in some respects not yet exhaustive, an intuitive interpretation comes from the mechanism of Bose-Einstein condensation, at least for helium 4. All liquids below a certain temperature transit into the solid phase; in fact, by lowering the temperature, atomic and molecular vibrations are reduced, allowing the formation of molecular bonds strong enough to trigger the transition to the solid phase. Helium is the only element that does not solidify at absolute zero unless a pressure of about 25 atmospheres is applied. This is due to the weak interaction of helium atoms even when thermal vibrations are minimized. Helium, being composed of an even number of nucleons (2 protons and 2 neutrons), has a total integer spin as it is the sum of an even number of half-integer spins. Therefore, for helium, the Pauli exclusion principle does not apply, consequently all atoms can occupy the same energy state and be described by a macroscopic wave function, implying a coherent collective behavior of all atoms, as happens in the Bose-Einstein condensates previously discussed.

In 1957, the superfluidity of helium 3, a rare isotope of helium 4 with a single neutron and two protons at a temperature of about 3 mK (3 thousandths of a degree above absolute zero) was discovered. This discovery confirmed that the phenomenon could not be attributed to a simple Bose-Einstein condensation. In fact, helium 3 is a fermion for which the Pauli exclusion principle applies and not the Bose-Einstein condensation. However, at a temperature close to absolute zero, a weak electromagnetic interaction can determine the formation of pairs of helium 3, which are bosons and the phenomenon can be attributed to one similar to that of helium 4 or superconductivity.

There are less known transitions and only recently discovered like those studied by physicists David Thouless, Duncan Haldane and Michael Kosterlitz, winners of the 2016 Nobel Prize in Physics for their research on the topological phases of matter and the related transitions that have allowed to identify new states of matter. These are transitions in very thin (two-dimensional) materials that below a certain temperature are characterized by the presence of pairs of strongly bound vortices which determine a distinctly quantum behavior of the material overall. The phase transitions we are talking about are not as simple as those ice-water and the explanation required the use of sophisticated mathematical models based on topology (a branch of mathematics that studies the properties of objects that do not change when they are deformed without tearing). Developed in the early 1980s, the theories of topological transitions have allowed to explain important macroscopic quantum phenomena such as the quantum Hall effect, discovered in 1980 by Klaus von Klitzing, awarded the Nobel Prize in Physics in 1985. The aforementioned effect involves the quantization of a material's conductivity as the applied magnetic field varies.

The three scientists soon realized that topological transitions allowed to predict and study structures of great theoretical and experimental interest such as magnetic atom chains and the most recent topological insulators, semiconductors and superconductors that represent a new state of matter as well as a new frontier of condensed matter physics. Particularly interesting are the topological insulators such as the *Quantum State Hall systems—QSH* which are insulators inside and, at the same time, conduct on the surface as they have surface states characterized by the propagation of electrons with aligned spins. These materials in addition to having an important and perhaps fundamental role for the development of elementary quantum bits can be used as a tool to investigate elusive elementary particles such as *axions* (possibly responsible for dark matter) or Majorana fermions, particles identical in every way to their antiparticles.

The extraordinary manifestations of quantum effects in condensed matter do not end here. In a material with very particular properties, it is even possible to witness quantum-relativistic phenomena, which, as such, are described by the Dirac equation. The material in question is graphene, which is an extremely thin graphite sheet (the common pencil for writing) consisting of a monoatomic layer of carbon atoms arranged in a honeycomb pattern (Fig. 3.5), and can be considered in all respects a two-dimensional material, in other words a portion of a plane whose height is zero.

Although already predicted at the end of the forties of the last century, graphene was only realized in 2004 by the English physicist Andrej Gejm and his doctoral student Konstantin Novosëlov, with a very simple technique based on the exfoliation of graphite using adhesive tape.

From an application point of view, graphene has enormous potential linked to its truly unique characteristics: extremely resistant, very light, flexible like

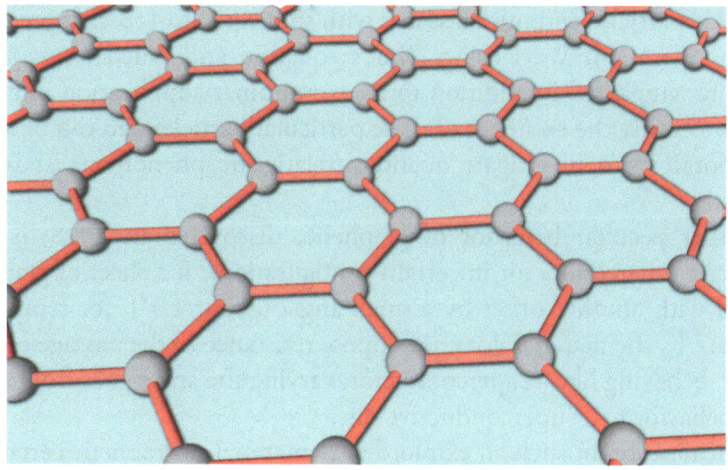

Fig. 3.5 Molecular model of graphene. Carbon atoms form a hexagonal cell structure

plastic, excellent conductor of heat and electricity and transparent to light. Its surface density is extremely low ($0.77 \ mg/m^2$), just over 5 g of graphene would be enough to cover a football field. Moreover, it is so resistant (100 times more than steel) that less than a thousandth of a gram can support the weight of about 4 kg.

These properties allow the use of graphene in applications of great interest such as electronics, sensors, detectors, optics, biomedicine, and wellness (tennis rackets, skis, helmets, bike frames and wheels). Considered a material of a thousand wonders, graphene, according to many scientists, will produce a real revolution in the science and technology of materials, in some aspects already started considering the numerous applications of this special material.

But the wonders of graphene also include truly peculiar quantum/relativistic effects, allowing the use of graphene to study phenomena that would require large particle accelerators.

In graphene, due to the interactions of carbon atoms arranged in a regular flat hexagonal structure (Fig. 3.5), a very unique band structure is formed, deeply different from the classic band structure of ordinary metals and semiconductors. The shape of the valence and conduction bands is similar to two cones, one inverted and joining at the tip (hourglass), essentially forming a structure without a gap, unlike insulators and semiconductors. This particular band structure gives electrons atypical characteristics: they are much faster than solid electrons reaching speeds of about 1000 km/s, they possess energy that depends on speed and not on the square of speed, they move in one direction and cross obstacles by tunnel effect. These properties are typical of relativistic particles without mass and with spin equal to 1/2, first predicted in 1929 by Hermann Weyl using Dirac's equation and called *Dirac fermions*. Therefore, graphene, in addition to its extraordinary application impact, has allowed to verify the existence of these particular particles and can be an effective laboratory to investigate quantum/relativistic phenomena at relatively low costs.

Another peculiar behavior of graphene, discovered in 2018, is that it becomes a superconductor in certain configurations: if a sheet of graphene is overlaid with another offset by a small angle (about 1.1°), for temperatures below 1.7 K, the material does not oppose resistance to the passage of electric current, behaving like a superconductor; varying the angle goes from an insulating behavior to a superconductive one.

The realization of such an extraordinary material as graphene certainly did not go unnoticed by the Royal Swedish Academy of Sciences, which in 2010 awarded the Nobel Prize in Physics to Gejm and Novosëlov for having realized the first sheets of graphene.

We conclude this section with the following observation: in the introduction of the famous article by Peter Higgs from 1964 in which the missing link of the standard model was predicted, namely the Higgs mechanism that allows particles to acquire mass, the author argues that the mechanism he was describing was the relativistic analogue of what the American physicist Philip Anderson had reported in an article on superconductors, in which he theorized that, in case of symmetry breaking, the excitations of the fields (the particles) acquired mass. In short, the study of a particular type of condensed matter, superconductors, had made Anderson intuit an important concept that certainly inspired Higgs.

At this point we have enough elements to believe that Pauli had made a somewhat hasty assessment in considering solid state physics a "dirty" physics.

At the end of this very brief compendium on condensed matter physics, it is necessary to clarify that it is a very large field and we have focused on the most relevant topics also taking into account their technological impact, but certainly other equally important topics have been neglected.

3.3   Quantum Mechanics in Everyday Life: The Technological Impact of Quantum Physics

In the previous chapter, we saw that thanks to relativistic corrections, the satellite navigator of our car or our smartphone allows us to reach even very distant destinations with an accuracy of a few meters. Yet, in the years when the theory was developed, no one would ever have thought that it would have had an impact in everyday life.

Quantum physics, in some ways even stranger and further from common sense than the theory of Relativity, has allowed the realization of applications that have entered in a capillary way in our daily life, to the point that it is difficult to imagine a day in which a person does not use a device or a system that is based on the principles of quantum physics. It is a real technological revolution that has changed our way of living on par with the electromagnetic revolution that brought us lighting, electric motors, radio and television.

In this section, we will try to review the most important applications of quantum physics, especially from a technological impact point of view. Naturally, as with physics of matter, many applications will not be mentioned not because they are less important, simply because it would be impossible to describe them all in a section of a book, whose purpose is to stimulate the reader to take a small look at this strange and wonderful world of modern physics.

We start this brief review with the application that is most familiar to us: semiconductor-based electronics. It is contained in all the devices we use from morning to night, think of smartphones, computers, televisions, the electronics of modern cars, etc. In each of these devices thousands or hundreds of thousands of transistors or other circuit elements based on semiconductors are used.

But at the base of the operation of a transistor there is the quantum theory of bands and the tunnel effect. In particular, this latter effect is at the base of the operation of some types of diodes known as tunnel effect diodes or Esaki from the name of their inventor. In this type of diode, the flow of electrons is regulated by the height of a potential barrier and is made up of electrons that cross the barrier due to the tunnel effect (see Sect. 2.2). This flow of electrons can be stopped very quickly by acting on the barrier and since this variation can be very rapid (even less than 5×10^{-12} seconds) this device is used when extremely quick responses are needed.

Transistors can be used both as signal amplifiers and to create complex digital electronics elements, such as memories or powerful microprocessors for calculation and data and/or image processing. There are two types of transistors: one is bipolar junction and the other is field effect. Even though the detailed description of these types of transistors is beyond the scope of this book, we will later have the opportunity to describe in more detail the *PN junction* which is the basis of transistor operation.

It is clear that since the first transistor prototype of 1948 there has been a significant technological evolution including the realization of integrated circuits capable of containing in a single silicon chip a huge number of transistors of both bipolar junction and field effect type. The first integrated circuit, made in 1958 by the American electrical engineer Jack St. Clair Kilby (Nobel Physics Prize in 2000), contained just ten components and marked the birth of microelectronics. In fact, with the development of techniques of increasingly sophisticated fabrication techniques, we have reached a very high level of integration allowing a miniaturization unimaginable until a few years ago. Currently, the most complex and advanced chip in the world is the one made by the American company "Nvidia". The H100 chip was made using advanced fabrication technology that allows structures of just 4 nm (4 billionths of a meter) and contains a whopping 80 billion transistors capable of performing a quadrillion operations per second.

Even though many times we refer to semiconductor-based electronics as classic electronics to distinguish it from the prototypical one based on superconducting circuits or quantum bits, the operating principles are based on semiconductor physics and therefore on quantum physics.

Another large-scale application of quantum physics is certainly the one related to LASER technology, an acronym for *Light Amplification by Stimulated Emission of Radiation* (amplification of light through stimulated emission of radiation). The first working prototype of a laser was made in 1960 by the electronic engineer and physicist Theodore Maiman at the Hughes Research Laboratories in Malibu, California, based on three fundamental elements: the concept of stimulated emission introduced by Einstein in 1917, its use to amplify radiation theorized by the Russian physicist Valentin Fabrikant in 1939 and above all the fundamental studies carried out in the mid-1950s by Charles Hard Townes, Nikolay Basov and Aleksandr Prokhorov who were awarded the Nobel Prize in Physics in 1964.

As can be inferred from the acronym, the laser is a device capable of amplifying light through a stimulation process of a substance which is called *active medium* and that determines the main characteristics of the laser. Active media can be solid (semiconductors, neodymium, ruby), gaseous (carbon dioxide, argon, fluorine, chlorine) or liquid (dye lasers, methanol, ethanol).

To understand the operating principle of a laser, remember that, as seen in Sect. 2.1, the energy levels of electrons in atoms are quantized and that, if an electron jumps from an orbit at higher energy to one at lower energy, the atom emits a photon whose frequency is given by the ratio between the difference in energies of the starting and arriving orbits and Planck's constant ($f = (E_f - E_i)/h$). The quantum leap of the electron and the consequent emission of the photon can be spontaneous or, as Einstein intuited, stimulated from the outside by another photon with the same energy as the one that will be emitted (Fig. 3.6).

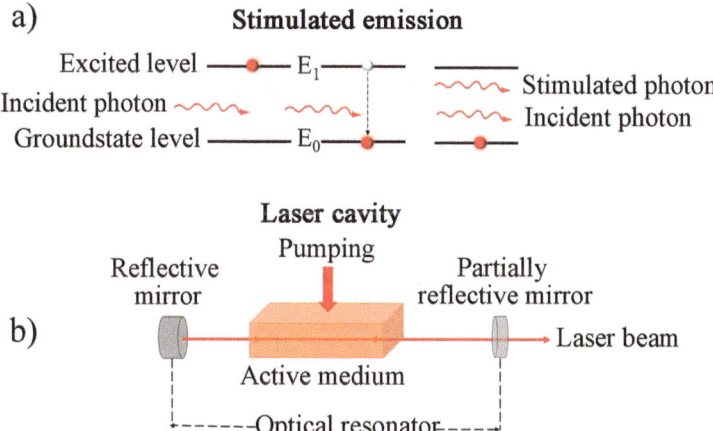

Fig. 3.6   (a) Representation of the stimulated emission process. (b) Basic schematic of a Laser operation: within a resonant cavity formed by two mirrors, the emission of photons is stimulated. The photons are then reflected by the mirrors and stimulate other photon emissions through a kind of chain process

But since electrons tend to occupy the lower energy states, to have a stimulated emission it is necessary to bring the electrons to the higher energy level, it is necessary to perform what is called *a population inversion*.

For this purpose, two techniques are mainly used: optical pumping and electronic excitation. In the case of optical pumping, an auxiliary light source is used to bring the electrons to the excited levels and is particularly used for ionic crystal lasers (ruby and neodymium).

In the case of electronic excitation, energetic electrons are injected into the active material and collide with the atoms in the ground state, which produces an excitation to the higher energy levels. Consider as an example an electrical discharge through a gas, a technique typically used for gas and semiconductor lasers. However, it should be clarified that in reality things are slightly more complicated than the simple scheme shown in Fig. 3.6. In fact, when the two levels have reached the same number of electrons, the number of absorbed photons is equal to that of the photons emitted by stimulation and the material effectively becomes transparent. To avoid this condition, three or four level systems are usually used in which the extreme levels (first and last) are used for absorption and the internal ones for stimulation. In the case of a three-level system, following the absorption of photons, the electrons move from the first to the third level (population inversion), and then quickly decay to the second level where they are stimulated to decay to the first level by other photons, having a different frequency from those used for the population inversion.

To amplify the light coming from the stimulated emission, however, it is essential to create a particular geometric structure called *resonant cavity or resonator*. The latter has the function of creating a kind of chain reaction in order to amplify the laser light. One of the simplest ways to create a resonant cavity is to insert the material, on which the population inversion (active material) will be performed, between two parallel mirrors as shown in Fig. 3.6. In fact, in this case, due to spontaneous emission, the electromagnetic wave (photon) propagates back and forth in the direction orthogonal to the mirrors, and thanks to the process of stimulated emission, photons are generated at each passage in the active material. At this point, if one of the two mirrors is made partially transparent, the useful beam of photons will come out of it.

Compared to light, lasers, by virtue of the operating principles on which they are based, possess very peculiar properties such as monochromaticity, directionality, and brilliance. A laser beam is an excellent approximation of a monochromatic electromagnetic wave, i.e., having a single oscillation frequency which in the case of the laser is $f = (E_f - E_i)/h$. With lasers, it is possible to obtain a beam of light with characteristics of monochromaticity significantly superior to those obtainable through the most monochromatic conventional sources such as spectral lamps. Since the material is placed in a resonant cavity made up of two mirrors, only an electromagnetic wave that propagates in the direction orthogonal to the mirrors can oscillate. This gives the beam of light extreme directionality, unlike the light from a normal incandescent bulb that emits light in all directions. Moreover, a laser beam at a great distance diverges minimally: a green beam from an Argon laser, with a starting section of one centimeter in diameter, widens to a section of three centimeters in diameter after a path of 500 m.

Finally, brilliance is defined as the power emitted per unit of surface and per unit of solid angle. Lasers have a very high brilliance, mainly due to the small value of the beam divergence and of course also the high power.

The first commercial application of a laser dates back to 1967 in Cambridge, England, when a laser beam was used to cut a steel sheet one millimeter thick, taking advantage of the high energy that lasers can focus on a very small area. This first application paved the way in the 1970s for the massive use of lasers in the automotive industry where lasers were used to cut and weld metals. Subsequently, smaller lasers were also used for the processing of plastic and rubber. Today, lasers are used in many fields, just think of barcode readers in supermarkets or CD and DVD readers. Even in the medical field, lasers are used with great success, particularly in surgery, dermatology, oncology for the ablation of superficial tumors, ophthalmology, otolaryngology.

A spectacular application of lasers is undoubtedly holography, a technique that allows the construction of three-dimensional images using beams of light. Although theorized in 1948 by Hungarian physicist Dennis Gabor (Nobel Prize for Physics in 1971), the technique began to be successful only from the early sixties using the first laser beams. Today, holography is extremely advanced and allows for high-definition three-dimensional images of very small structures such as red blood cells, whose morphology provides valuable information from a medical point of view.

Finally, from a research point of view, the laser is a fundamental tool for the study of some branches of physics such as photonics, classical and quantum optics, photonic quantum computer, inertial thermonuclear fusion.

Before moving on to another application, it is worth remembering that there are advanced laser systems used to study very fast phenomena such as the dynamics of electrons in matter or in atomic and molecular systems. These are sophisticated experimental devices capable of emitting very short ultraviolet light pulses that can reach up to a few attoseconds (10^{-18} s). To realize how short an event lasting an attosecond is, consider that one second is equivalent to 1 billion billion attoseconds and the age of the Universe is instead about half a billion billion seconds; therefore, on a time scale the distance between an attosecond and a second is the same as that between a second and the age of the Universe!

The importance of these tools that allow to "photograph" very rapid phenomena has not escaped the experts of the Nobel Prize commission, who awarded in 2023 the prestigious prize for physics to Pierre Agostini, Ferenc Krausz and Anne L'Huillier for their fundamental contribution to the development of the aforementioned attosecond laser systems.

To avoid misunderstandings, it is necessary to clarify that these exceptional lasers do not allow to film the trajectory of the electrons as this is not possible both due to the wave nature of electrons and the uncertainty principle. Indeed, knowing the trajectory of a particle implies knowledge at every moment of the position and speed, which is not possible due to the aforementioned uncertainty principle. However, it is possible to study ultrafast processes such as those related to electron transitions between different quantum states or to interaction phenomena between the same electrons, which are very useful for understanding fundamental aspects of atomic physics, condensed matter physics and molecular chemistry. Such studies will, most likely, have a significant impact on materials science as well as on technologies (electronics, quantum computing) and on biomedicine, where it will be possible to develop advanced in vitro diagnostic techniques on blood samples.

The operating principle of lasers introduces us to another disruptive application of quantum physics. Indeed, the light emitted following the transition of quantized energy levels is also at the base of modern lighting. Now, most of the bulbs in our homes, in public places, on the streets and even the lighting of TV screens, PCs and mobile phones, are based on LEDs, an acronym for *Light Emission Diode*.

A LED is made up of two adjacent or overlapping semiconductors, with different electrical properties, one with a contamination of atoms ready to give up electrons known as *n*-type semiconductors and the other with atoms ready to capture electrons creating positive charges (*holes*), *p*-type semiconductors. This structure is called a *PN* junction and the operation of contaminating the two superconductors is called *doping p*-type or *n*-type for the two semiconductors. The doping is achieved by inserting into the semiconductor *p* atoms acceptors of electrons such as boron, aluminum or gallium, which having only three electrons in the outermost orbital tend to share one of the four outer electrons of silicon, forming a positive charge bond due to the partial absence of an electron from the neighboring silicon atom. Therefore, the *p*-doped semiconductor has an excess of positive charges due to these absences of electrons. Similarly, for the semiconductor *n*, an electron donor atom like phosphorus, arsenic or antimony is used, which, having five electrons in the outermost orbital tend to partially give one of the five electrons to the neighboring silicon atom, producing an excess of negative charges.

At the interface between the two *n* and *p* semiconductors, the excess electrons tend to diffuse from the *n* semiconductor to the *p* one, leaving a positive charge zone in the *n* semiconductor and forming a negative charge zone in the *p* one. The process at a certain point stops as the accumulation of negative charges near the interface opposes the diffusion of electrons coming from the *n* semiconductor. The region formed near the contact area between the two semiconductors is called the *depletion/space charge or also active region*, it has a size of a few thousandths of a millimeter and is characterized by the presence of a voltage (0.5–0.6 V for silicon and 0.2 V for germanium) due to the distribution of positive and negative electric charges formed at the interface of the two semiconductors.

Returning to the operation of an LED, the application of an appropriate electric voltage to the ends of the diode causes the migration of electrons or holes, present respectively in the conduction and valence bands, towards the active zone. As stated in the previous section, the two bands are energetically separated by a gap. During migration in the active zone, the electrons move from the conduction band to the valence band to recombine with the holes, emitting photons with a frequency in the visible spectrum (Fig. 3.7). The

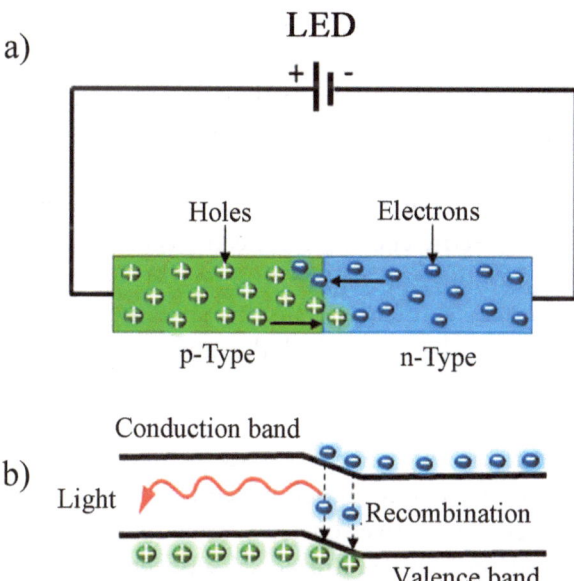

Fig. 3.7 **(a)** LED (light-emitting diode) operating scheme. The voltage applied to the ends of the two *p* and *n* semiconductors generates a flow of electrons that near the interface of the two semiconductors move from the conduction band to the valence band **(b)**, generating light whose frequency is proportional to the energy difference between the two bands

semiconductors most commonly used for the production of LED lamps are gallium arsenide, gallium phosphide, silicon carbide, and gallium and indium nitride.

The great advantage of LED lamps is energy consumption: for the same level of illumination, it is extremely lower compared to traditional lamps. Considering that about a quarter of the world's electricity consumption is due to lighting, this innovative lighting technology also has a significant impact from an ecological point of view.

Furthermore, LEDs have a lifespan ten times longer than fluorescent lamps, they do not emit ultraviolet light, which is present in neon lamps, nor infrared light causing heating. The evident technological and ecological impact of this type of lamp has revolutionized lighting, allowing the three Japanese researchers (Isamu Akasaki, Hiroshi Amano, and Shuji Nakamura) to win the Nobel Prize in 2014 for the creation of the first blue LED in 1992.

The creation of the blue LED was crucial for the development of large-scale white LED lighting. In fact, green and red LEDs had already been created in the Sixties but to obtain white light, the blue light diode was needed, which combined with the other two produced white light. The combination of the three different diodes allowed to combine the three fundamental monochromatic lights (red, green, and blue) to obtain white light or any other color. Alternatively, white light can be obtained using a blue LED with a phosphor coating to convert the blue light into white light through the process of fluorescence.

Let's now move on to two very widespread and important applications. Based on the photoelectric effect, photoelectric cells and photovoltaic panels have occupied a place of great importance in our daily life.

Every gate, door, or shutter, operated by an electric motor are equipped with photoelectric cells thanks to which the moving object stops and typically goes back, avoiding unpleasant accidents or long waits. Photoelectric cells also allow the automatic opening of doors in offices, shops, or supermarkets and are also used in alarms and anti-theft devices. The operating mechanism is based on the use of a vacuum tube in which there are a cathode at negative voltage and an anode at positive voltage. A beam of light with an appropriate frequency to ensure the photoelectric effect (see Sect. 2.1), strikes the cathode producing photoelectrons attracted by the anode and thus producing a photo-induced electric current. If something (an object, a person) interposes between the light source and the cathode, the latter no longer emits electrons and the current is interrupted; in this case the resulting absence or decrease in current, triggers an electronic device that depending on the application provides appropriate commands such as gate opening/blocking, automatic door opening, alarm activation or alarm, etc. Currently, vacuum tube photoelectric cells are almost entirely replaced by those semiconductor in which light produces an electric voltage at the ends of a silicon crystal. The interruption of the circuit in this case causes a sudden drop in voltage rather than current as in the case of the vacuum tube.

This type of photocell is also widely used for the realization of photovoltaic panels that transform solar energy into electric energy, of fundamental importance for the ecological transition.

The operating principle of a photovoltaic panel is based on a very similar effect to the photoelectric effect, namely the photovoltaic effect discovered in 1839 by Alexandre Edmond Becquerel who observed that, exposing particular electrodes in a conductive solution to sunlight, they generated a small current flow. It took many years, however, to create the first silicon-based photovoltaic cell which took place in 1954 at the Bell labs in the United States. This first prototype produced a modest amount of energy, it was able to power just a small transceiver, but laid the foundations of the new photovoltaic technology.

Similarly to the photoelectric effect, when a semiconductor material (typically silicon) is hit by light radiation, the electrons absorb the light energy and move from the valence band to the conduction band, with the difference that in semiconductors the electrons do not escape from the material but move into the conduction band with a higher energy.

The basic element of a photovoltaic panel is the single photovoltaic cell very similar to the LED but with an inverse mechanism. In the case of photovoltaic cells, the incident light produces a current and then a voltage at its ends, converting the energy of the photons into electric energy unlike LEDs, where the current produces the emission of light and therefore light energy.

In particular, when light hits the *n-type* semiconductor, where there is a greater quantity of atoms arranged to give an electron, electrons are generated that move into the conduction band and, driven by the voltage at the ends of the PN junction, flow in a closed circuit generating current (Fig. 3.8).

The efficiency of converting solar energy into electrical energy in recent years has increased significantly reaching 20%, that is, of all the solar energy collected by the panel, 20% is converted into electrical energy. This allows to meet the energy needs of an average family (about 3 kW) with a surface of about 30 m² of photovoltaic panels installed on the roof of the house.

Now we will talk about applications of quantum physics less familiar and widespread but equally important, starting from those in medical field.

A significant step forward in the progress of medicine and surgery has undoubtedly been possible thanks to both diagnostic imaging, which has allowed for reliable and precise diagnoses, and therapeutic techniques based

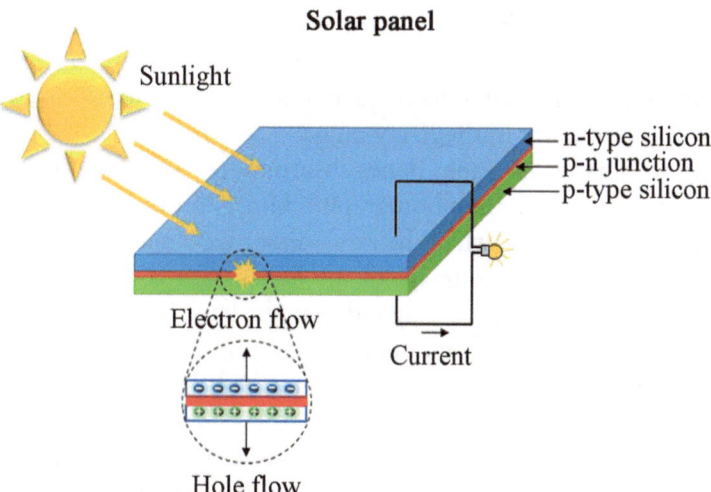

Fig. 3.8 Operating scheme of a solar panel. Sunlight produces a flow of electrons that flowing in a closed circuit generates a current

on the use of radiation and radiopharmaceuticals. Based on principles or phenomena of quantum physics, this important field of medicine is called *nuclear medicine*.

Before discussing the most current and perhaps most interesting diagnostic techniques from a quantum physics perspective, it is worth remembering that the discovery of X-rays by the German physicist Wilhelm Conrad Röntgen in 1895, immediately brought about a real revolution in the medical field because the German physicist, realizing the value of his discovery, refused to patent it to speed up its application. In fact, just one year after the discovery of X-rays, the first radiographs were already being taken on the battlefield. For this revolutionary discovery, Röentgen won the first prestigious Nobel Prize established for the first time in 1901. We briefly recall that X-rays are electromagnetic waves with a very small wavelength, between 1 *nm* (10^{-9} m) and 1 *pm* (10^{-12} m) and for this reason, they are very penetrating. They pass through the human body and can impress a photographic film placed behind the body. They are produced by bombarding a metal plate with a high-energy electron beam obtained by applying a voltage even higher than 25,000 volts between the cathode that emits the electrons and the metallic anode. Since an electric charge subjected to acceleration or deceleration emits electromagnetic waves, the abrupt slowdown of the electron beam following the collision with the metal plate produces high-frequency electromagnetic radiation and therefore a short wavelength. Alternatively, the collision of the electron beam can cause the expulsion of an electron belonging to the innermost orbitals of the atoms of the plate; the subsequent decay of an outer electron to the level of the expelled electron produces a photon whose energy, being equal to the difference in energy between levels very far apart, is very high. Since the energy of a photon is proportional to its frequency, it is possible to obtain in this way photons with frequency in the X-ray range.

During an X-ray radiograph, bones absorb more X-rays compared to muscles and skin; therefore, the photographic film is more exposed in the area where there are no bones, while the remaining part remains clear as it is little exposed and reproduces the image of the bones with good definition. From the first radiological images, X-ray radiographs have been significantly optimized and improved up to computed tomography (CT) invented in 1971 by the young English engineer, Godfrey Hounsfield. CT, thanks to the aid of computerized reconstruction techniques, allows obtaining three-dimensional images of various tissues. Using significant doses of X-rays, CT remains an invasive examination and as such cannot be performed frequently.

In the eighties, another revolutionary diagnostic technique was born based on a quantum mechanics phenomenon known since the end of the forties:

nuclear magnetic resonance (NMR). Not using X-rays or ionizing radiation, NMR is a non-invasive technique with huge potential from a diagnostic point of view. The operating principle of NMR is based on the phenomenon of nuclear magnetic resonance observed independently for the first time in 1946 by physicists Felix Bloch and Edward Purcell awarded the Nobel Prize for physics in 1952.

As reported in Sect. 2.1, all the constituents of stable matter have spin, a sort of intrinsic angular momentum which implies, in the case of a charged particle, also a magnetic moment. In other words, a proton behaves like a microscopic magnet. We know that if we put a magnet in an external magnetic field, like the needle of a compass in the Earth's magnetic field, the magnet tends to align with the external magnetic field. In fact, the compass needle always points to the magnetic north as it aligns with the Earth's magnetic field. In the presence of a static magnetic field (polarization field), the protons in the body and therefore essentially all the protons of the water contained in our body tend to align with the external magnetic field. However, just as a spinning top subjected to the Earth's gravitational field rotates both around its own axis and around the direction of the gravitational field (precession motion), in the same way the magnetic moment M of the proton performs a precession motion around the direction of the external magnetic field, whose frequency (called Larmor) depends on the intensity of the field, and the type of atom to which the proton belongs. In the case of hydrogen, the most common element in the human body, the Larmor frequency is 42.58 MHz for each applied tesla. In NMR the applied polarization magnetic fields vary from 1.5 to 3.0 tesla (1 tesla is equal to 20,000 times the Earth's magnetic field) which corresponds to a Larmor frequency in the radio frequency field (60–150 MHz).

Such high magnetic fields are necessary, as the signal/noise ratio that provides the quality of the signal and therefore also of the images is directly proportional to the applied static magnetic field.

If we now apply an electromagnetic pulse at radio frequency (excitation field) even of modest intensity but with a frequency equal to that of Larmor and orthogonal to the static field of polarization, it induces the flipping of the magnetic moment of the protons M_z in the plane orthogonal to the static field, or in a plane that forms any angle with the plane containing the static field depending on the duration of the magnetic pulse at radio frequency (Fig. 3.9). It's as if the magnetic needle of the compass rotated 90° from the initial position where it indicated the north. Once the radio frequency magnetic radiation is removed, the protonic magnets tend to return to their initial position and in doing so induce an electric voltage in the detection coils through the well-known phenomenon of electromagnetic induction.

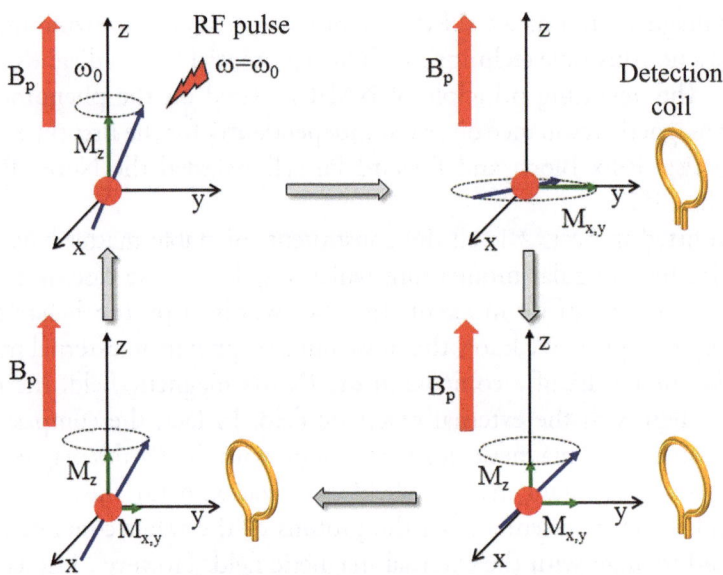

Fig. 3.9 Basic scheme of nuclear magnetic resonance. The magnetic moment of the protons performs a precession motion around the direction of the static polarization field with the Larmor frequency ω_0. An electromagnetic pulse at radiofrequency (RF) having a frequency ω_0 produces a flipping of the magnetization M_z in the x-y plane (M_{xy}). Once the pulse is removed, the system will return to the initial state with characteristic times that depend on the tissue being analyzed. The electrical signals induced in the detection coils during the return to the initial condition are processed to construct the anatomical-morphological images

It is precisely these signals detected by the coils that, properly processed, provide the image of the organ or tissue that you want to analyze. In order for there to be a measurable signal, a considerable number of protons are needed that after the removal of the excitation field return to the initial position, this process is also known as *relaxation*. The amplitude of the signal induced by relaxation depends on the number of protons while the time taken to return to the initial state depends on the chemical-physical characteristics of the cell cluster or portion of tissue that is being analyzed. In particular, there are two types of relaxation: the first is due to M_z that goes from zero to its initial value and the second is due to the flipped component (M_{xy}) that goes from its maximum value to zero (Fig. 3.9). The time taken by M_z to return to its initial value, known as *T1*, is greater than the one that takes M_{xy} to return to zero (*T2*), even though apparently the two times should be equal. In reality, the magnetic moment of the protons flipped in the x-y plane get out of phase quickly and cancel each other out, resulting in a magnetization M_{xy} null before M_z has returned to its maximum value.

From a clinical point of view, *T1* is related to the anatomy of soft and fatty tissues while *T2* is related to fluids and to pathological conditions such as tumors, traumas or inflammations.

In addition to the static magnetic field and the radio frequency one, to distinguish and identify the signals coming from the various areas of the structure analyzed, the use of magnetic field gradients or spatially variable magnetic fields with different intensities depending on the point considered is necessary. These small intensity fields add to the static field determining a different Larmor frequency for each point considered. In this way, when the excitation field is applied with a precise frequency, only the areas characterized by the Larmor frequency corresponding to the field of polarization at a certain point are excited. By changing the frequency of excitation, the point at which it is equal to the Larmor frequency changes and this allows for a complete scan of the area of the body to be investigated without moving the patient.

The use of magnetic field gradients, first introduced in 1972 by the American chemist Paul Lauterbur, allowed for obtaining the first two-dimensional and three-dimensional images. The use of particular mathematical algorithms, subsequently developed by the English physicist Peter Mansfield, allowed to process the data and produce high-quality images in a few seconds instead of several hours.

For this fundamental step forward in the use of the phenomenon of nuclear magnetic resonance in the biomedical field, the two scientists received the Nobel Prize for Medicine in 2002. The static magnetic fields, the radio frequency one and the magnetic field gradients are generated by particular coils that surround the bed on which the patient is positioned.

This diagnostic technique, considered much less invasive compared to CT for the absence of ionizing radiation, has the advantage of providing high-quality images also for soft tissues as well as varying the contrast of the images greatly. Naturally, images of tissues less rich in water like bones show a lower quality; therefore, in these cases, it is preferable to use CT.

NMR is used for diagnostic purposes mainly for diseases affecting the central nervous system, joint ligaments, the cardio-circulatory system and cartilages, providing extremely useful images for diagnosing the presence and/or progression of tumors or any morphological/anatomical malformation.

Using very intense magnetic fields, this investigation is contraindicated for wearers of metal prostheses or pacemakers, moreover, since the vibrations of the coils used for the field gradients, due to the interaction with the static polarization field, generate very intense noises at the threshold of pain (about 120 dB), it is necessary to wear ear protection headphones during the examination, which lasts from 20 to 30 min.

In recent years, NMR systems with a static polarization field of 7 tesla have been developed, which allows for obtaining images of very high quality and better spatial resolution or greater ability to distinguish two contiguous areas. These new systems allow for better distinguishing of brain lesions from multiple sclerosis, to make early diagnoses in case of diseases neurodegenerative and more generally to observe particular anatomical-morphological, practically invisible with standard systems.

Another very useful diagnostic technique especially for oncological and neurodegenerative diseases is positron emission tomography (PET—Positron Emission Tomography). Developed by two physicists and a radiobiologist in the context of the development of medical applications using CERN technologies, PET was first presented in 1977 showing an image of a mouse's skeleton. The principle of operation of such diagnostics is based on the annihilation between electrons and positrons emitted following the decay of particular radioactive elements. In particular, a radiopharmaceutical is injected into the patient intravenously, consisting in a glucose solution to which a radionuclide is attached that, decaying, emits positrons. These annihilate almost instantly with the electrons of the surrounding tissues transforming into energy in the form of two gamma rays that, after having passed through all the surrounding tissues, are detected by suitable gamma ray sensors. Since tumor cells are particularly avid for glucose, the distribution of the detected gamma rays provides information on the quantity and position of tumor cells. Since the brain also mainly feeds on oxygen and glucose, glucose metabolic dysfunctions at the brain level provide important information for neurodegenerative diseases like Alzheimer's syndrome. In fact, PET is often used for the initial diagnosis of these brain diseases.

The gamma ray sensors are placed in a ring that surrounds the patient in order to detect the pairs of gamma rays that, for the conservation of momentum, move away in opposite directions forming an angle of 180° between them (Fig. 3.10). The two photons typically traverse different paths in the tissue before being detected; therefore, from the two measures of different attenuation, it is possible to trace back to the point where the pair of photons was generated. The images obtained provide information on the activities of the cells, attributing a more or less bright color depending on the intensity of metabolic activity. By moving the bed on which the patient is positioned inside the ring containing the sensors it is possible to perform a scan and produce a series of two-dimensional images corresponding to a section (slice) of the organ or structure under examination. Using appropriate software, the images related to the individual sections are combined and a three-dimensional image is obtained.

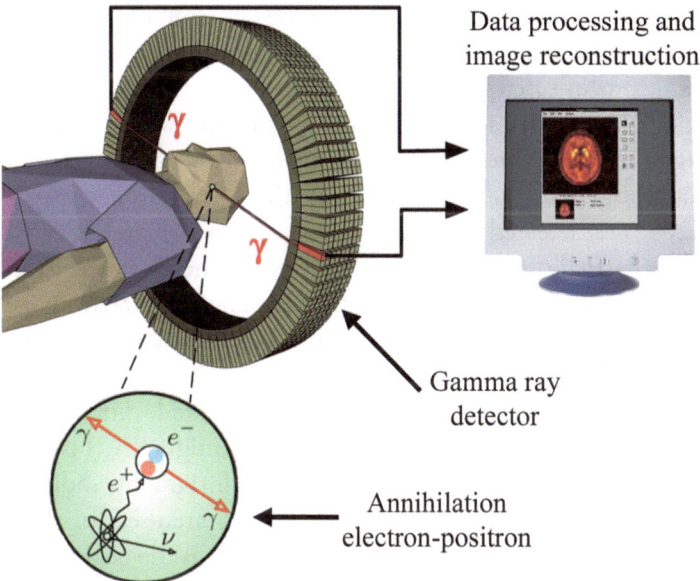

Data processing and
image reconstruction

Gamma ray
detector

Annihilation
electron-positron

Fig. 3.10 Schematic of positron emission tomography (PET). The electron-positron pairs produced by the decay of the radionuclide injected into the patient annihilate, generating two gamma rays detected by suitable sensors. Since the radionuclide is carried by a glucose-based solution, the examination identifies the areas where the glucose has accumulated. As tumors are very avid for glucose, the areas with the greatest accumulation of glucose typically correspond to cancerous neoplasms

Typically the radiopharmaceutical used in PET is fluorodeoxyglucose (FDG), chemically very similar to glucose in which a hydroxyl group (OH^-) is replaced with a radioactive fluorine atom (fluorine 18) (remember that the number following the element refers to the atomic mass number, that is the sum of the protons and the neutrons inside the nucleus).

Since the half-life of radioactive fluorine is about 2 h, a short time after the end of the examination the quantity of radioactive fluorine is negligible.

In the oncological field, PET also provides information related to the aggressiveness of the oncological pathology, to the effectiveness of chemo and radiotherapies and to the presence of any metastases.

Since it is a functional diagnostic in the sense that it highlights the functioning/physiology of an organ or tissue and not the morphology and anatomy, sometimes PET is combined with CT to overlay the morphological-anatomical image with the functional one provided by PET.

Other very useful diagnostic techniques similar to PET are scintigraphy and its three-dimensional version, namely computed tomography from single photon emission (SPECT—Single Photon Emission Computer Tomography).

Unlike PET, these diagnostics are based on the emission of gamma rays rather than positrons while the rest of the operating principle is the same. The radiopharmaceutical is injected into the patient and after a waiting time necessary for the radiopharmaceutical to reach the organ or the target tissue, the examination begins during which the gamma rays produced by the decay of the radionuclide bound to the radiopharmaceutical, typically technetium 99 or thallium 201, are detected by a *gamma camera*. In the case of SPECT, data acquisition is carried out by rotating the detection heads of the gamma camera around the patient's body. At each different angle, a planar image called a projection is acquired; the set of such projections then allows to obtain three-dimensional information.

Depending on the organ or tissue to be investigated, different radiopharmaceuticals are used. For example, for bone scintigraphy, bisphosphonate marked with technetium is used, or for thyroid scintigraphy, sodium pertechnetate with technetium. Scintigraphy and SPECT are used for the diagnosis of diseases and tumors affecting various organs such as the heart, bones, brain, liver, lungs, and kidneys.

The applications of modern physics to medicine are not limited only to diagnostics, but also to therapy. Beyond the already mentioned therapeutic applications of the laser and the well-known radiotherapy based on the use of X-rays aimed at the destruction of tumor masses, it is certainly worth mentioning oncological therapies through *hadrontherapy* and radiopharmaceuticals. Having become part of the essential therapies provided by the national health system in 2017, hadrontherapy uses collimated beams of protons or carbon 11 at high energy and is used in cases where ordinary radiotherapy is ineffective. The particle beams are accelerated through a particle accelerator to reach speeds of over 60,000 km/s in order to reach the energy sufficient to permanently destroy the tumor cells eliminating the possibility that they can regenerate. The definitive destruction of cells occurs due to the multiple breaking of chemical bonds within the DNA of the tumor cell preventing its reproduction, a condition not always guaranteed by radiotherapy. Furthermore, hadrontherapy allows to be even more selective, as, being very fast, the particles of the beam interact little with the superficial tissues and release most of their energy when they stop, a physical phenomenon known as the *Bragg peak*. It is therefore possible to calibrate the energy of the beam so that the protons or carbon atoms stop right at the desired point.

Radiopharmaceuticals, in addition to having a diagnostic function, can to also be used for therapeutic purposes. In fact, once the target is reached thanks to the *carrier* molecule, the radioactive atoms emit gamma radiation or beta radiation which, if properly dosed, can destroy the tumor cells present in the immediate vicinity of the radiopharmaceutical. The main advantage of this technique is the high selectivity due to the precision with which the radio-pharmaceuticals reach and bind to the tumor cells. Excellent results have been obtained for thyroid cancer using iodine 131 as a radionuclide to eliminate, downstream of the surgical intervention, any residual tumors or metastases in the neck lymph nodes. However, these therapies are proving effective also for other types of tumors such as those of the bladder, prostate, gastrointestinal-pancreatic tract, and non-Hodgkin lymphomas.

Finally, speaking of radionuclides, we cannot fail to mention another application, not medical, but of extreme importance: the dating with *carbon 14*. Developed in 1947 by the American physicist Willard Frank Libby (Nobel Prize for Chemistry in 1960), this technique represents a powerful investigative tool in many sectors such as archaeology, paleontology, hydrology, oceanography, geology, and biomedicine. Carbon 14 is a carbon isotope with 6 protons and 8 neutrons and is present in the atmosphere in a very small percentage compared to carbon 12 consisting of 6 protons and 6 neutrons. It is estimated that there is one atom of carbon 14 for every 10^{12} (one trillion) carbon atoms. This very rare isotope forms in the atmosphere following collisions between cosmic ray neutrons and nitrogen 14 which represents about 80% of the Earth's atmosphere. Transformed into carbon dioxide, carbon 14 is assimilated by plants and therefore also by animals that feed on plants as well as by animals and humans who feed on animals and plants.

From its formation, carbon 14, being weakly radioactive, transforms into nitrogen 14 through the transformation of a neutron into a proton and the consequent emission of an electron (beta decay, see Sect. 2.4). The half-life, or the time for the initial number of carbon atoms to halve is about 5700 years. Since carbon 14 is continuously produced and absorbed with equal continuity by plants and animals in the form of carbon dioxide, a dynamic equilibrium is created thanks to which the amount of carbon 14 present in plants, animals, and humans is constant. The absorption process stops following death and the amount of carbon 14 begins to decrease slowly. In particular, it reduces by half after 5700 years, by a quarter after 11,400 years, by an eighth after 17,100 years and so on. After about 50,000 years it disappears completely. The dating technique is based on measuring the amount of residual carbon 14 in the sample to be analyzed. Based on the residual percentage measured, it is possible to trace back to the last time there was absorption of carbon 14 and therefore to the age of the object under examination.

It is obvious that this technique is essentially applicable for organic or inorganic samples that contain carbon. In the case where the amount of carbon 14 in a sample is null or below the sensitivity of the detection instruments, it is deduced that the sample is older than 50,000 years like the case of dinosaur bones.

The most famous measurement in terms of media resonance made through the carbon 14 dating technique was that performed on April 21, 1988 on the holy Shroud, the most important relic of Christianity. The outcome of the measurement, published in the prestigious *Nature* journal, attributed to the Shroud an age dating back to the late Middle Ages (1260–1390) or not to the first century AD. In the following years, several hypotheses have been put forward that have highlighted potential factors that would have cast doubt on the reliability of the measurement. In particular, the numerous contaminations and manipulations suffered by the holy Shroud over the centuries would have undermined the reliability of the results. To the great satisfaction of the Clergy, a study conducted in 2022 by the Institute of Crystallography of the Italian National Research Council has shown that the measurements made on a sample of the Shroud with a technique based on the use of X-rays are compatible with a dating of the relics of about 2000 years.

In order not to lose sight of the quantum aspect of these applications, we recall that radioactive decay processes (see Sect. 2.3) are purely quantum phenomena governed by weak interaction and explained within the framework of the electroweak quantum theory.

Other applications that are certainly worth talking about are those related to superconductivity described in Sect. 2.4. When thinking about superconductivity and therefore the total absence of resistance to the passage of electric current, the first application that comes to mind is the creation of superconducting cables for the transport of electrical energy without any dissipation due to the Joule effect. Indeed, the production of superconducting electrical cables represents an important application of superconductivity but not so much to replace the normal cables of the electrical network, which is quite impractical due to the need to use cryogenic liquids, but to create powerful electromagnets capable of generating magnetic fields of even a dozen tesla. An electromagnet is essentially made up of a coil of low resistivity wire: the current passing through the turns of the coil generates the magnetic field which is directly proportional to the intensity of the current. One of the problems associated with the creation of powerful electromagnets is the dissipated power that inevitably leads to the melting of the wire if the current is excessive. The use of a superconducting cable completely solves the problem as, having zero resistance, it does not dissipate energy due to the Joule effect and therefore does not heat up. Usually multi-filament cables of a superconducting alloy made of niobium and tin ($Nb_3 Sn$) or niobium and titanium (Nb-Ti), whose transition temperature is respectively 18.3 K and 9.2 K, are used.

The superconducting magnets thus created are applied in NMRs which, as seen, require fields up to 7 tesla and in large particle accelerators like the CERN in Geneva, where high intensity magnetic fields (8–10 tesla) are used to deflect and collimate the beams of elementary particles.

The superconducting magnets of CERN are cooled with superfluid helium at a temperature of about −271 °C (1.9 K) and formed by individual blocks that are joined with special welds for over 18 km, equivalent to 66% of the entire 27 km ring. The current flowing in the cables to generate such intense fields is about 8700 amperes. It is easy to imagine the extreme attention required during the design and assembly phase of the super-magnet in order to avoid any unwanted dissipation of energy. In fact, in the presence of such a high current, the transition to normal metal even of a small portion of the superconducting cable would produce a huge dissipation of energy causing serious damage. This phenomenon known as *quenching* is not rare in particle accelerators; in fact, during the testing phase of the LHC (*Linear Hadron Collider*) at CERN in 2008, a faulty interconnection caused the quenching of the superconducting magnet causing a serious accident.

Powerful superconducting magnets are also used for the construction of large reactors for nuclear fusion with the magnetic confinement technique in which, as seen in Sect. 1.4, powerful magnetic fields confine the plasma of tritium and deuterium in a toroidal geometry. The ITER reactor (see Sect. 1.2) provides for a superconducting magnet powered by a current of 68,000 amperes capable of producing a magnetic field of about 12 tesla.

Another striking application of superconducting magnets is related to the possibility of achieving magnetic levitation due to the repulsion of equal magnetic poles. Taking two magnets, we know from experience, that they attract or repel each other depending on how we bring them closer together. As with electric charges, in magnetism equal poles repel and opposite poles attract. The non-trivial difference is that in the case of magnetism we cannot isolate the magnetic pole as we do for electric charges. By creating repelling magnetic poles through powerful electromagnets, it is therefore possible to create magnetic levitation transport systems with the obvious advantage of minimizing friction and therefore gaining speed and fuel consumption.

An example is the magnetic levitation train that travels without touching the rails known as MAGLEV (abbreviation and contraction of magnetic and levitation), which uses powerful superconducting electromagnets. First implemented in Japan, the latest generation superconducting MAGLEV train uses superconducting electromagnets placed on the train and has reached a maximum speed of 603 km/h in the testing phase, surpassing all previous records. When stationary or at speeds below 150 km/h, the train rests on rubber wheels while at high speeds the superconducting magnets under the train interact with the electromagnets placed on the tracks and the train rises from the rails by about 10 cm. The motion occurs thanks to the thrust of the same electromagnetic forces that lift it. The train's braking is achieved by reversing the polarity of the electromagnets and using air brakes based on air friction.

The race for the super-fast train also involves other industrialized countries. In particular, China aims to build a MAGLEV train that would cover the Beijing-Shanghai (1300 km) stretch in just 2 h and 30 min. Currently, there is already a MAGLEV train in China on a short 30 km stretch that connects Pudong Airport (Shanghai) with downtown Shanghai in about 7 min.

Equally important and interesting are the small-scale applications of superconductivity, especially those based on the Josephson effect. Theorized in 1962 by Welsh physicist Brian Josephson and first observed in 1963 by Norwegian physicist Ivar Giaever (both Nobel laureates in physics in 1973), the Josephson effect predicts the tunnel effect of Cooper pairs through a classically forbidden barrier. So a tunnel not of the single electron but of a pair in which the distance between the electrons can be several hundred nanometers. This effect occurs when two superconductors are separated by a very thin barrier of an insulating material (about 1 nm). A structure of this kind is called a Josephson junction. If we provide current to the junction, no voltage is observed at its ends up to a certain current value beyond which the device switches in a very short time (10^{-12} s) to a state with finite voltage. The current that does not produce a voltage drop at the ends of the junction is due to the tunnel effect of the Cooper pairs, and is called the Josephson or supercurrent, while the threshold value beyond which the junction switches is called the *critical Josephson current*. Given these characteristics, one of the first things that came to mind for scientists was to use these devices to create very fast digital electronics using as bit *0*, the zero state at zero voltage, when the current is below the critical threshold, and as bit *1*, the state at non-zero voltage, for currents greater than the critical one. Indeed, starting from the mid-1970s, this type of superconductive electronics based on Josephson junctions was developed. The widespread use of this highly performant digital electronics was clearly limited by the need to cool the superconductors, making their use disadvantageous compared to semiconductor-based electronics, whose performance was constantly improving.

However, there are several applications where the performance of superconductive devices is so superior to other devices that the inconvenience of using cryogenic liquids or complex cooling systems is overshadowed. One of these applications certainly involves the use of superconductive quantum interference devices, consisting of two Josephson junctions placed in a superconductive ring. These devices, better known as SQUIDs (acronym for Superconducting Quantum Interference Devices), are the best existing magnetic field sensors with a sensitivity such that they can measure the magnetic fields produced by neuronal currents. They can measure magnetic fields that are 50 billion times smaller than the Earth's magnetic field. Thanks to this enormous sensitivity, these devices are used in various fields such as magnetic microscopy, non-destructive material analysis, geophysics, astrophysics, nanomagnetism, quantum computing, and biomedicine.

The most important application of SQUIDs is in the biomedical field, particularly in neurology. Large systems containing several hundred SQUID sensors are used for magnetoencephalography (MEG) studies, a completely non-invasive technique aimed at measuring the magnetic field generated by currents flowing in neurons instead of the electric potentials that are instead measured by electroencephalography (EEG).

MEG systems consist of a thermally super-insulated cylindrical container (*dewar*), inside which the SQUID sensors are placed on a helmet-shaped support to fit the shape of the patient's head (Fig. 3.11); the *dewar* is filled with liquid helium to cool the sensors. The distance between the sensors at a temperature of -269 °C and the patient's head is just 2 cm. The super-insulation ensures that the temperature on the outer surface of the *dewar* is equal to that of the environment. The entire system is placed in a high electromagnetic shielding cabin to prevent environmental magnetic signals, which are much stronger than the signals generated by the brain, from disturbing the measurement by completely covering the magnetic signal that is to be measured. Unlike MRI, magnetoencephalography systems provide extremely useful functional images for both basic neuroscience studies and clinical applications such as identifying epileptic foci and mapping brain areas before surgical interventions to make the procedure less invasive. Also of considerable interest in the clinical field is the use of MEG for the study of neurodegenerative diseases (Alzheimer's syndrome, Parkinson's disease, amyotrophic lateral sclerosis, frontotemporal dementia). Compared to EEG, MEG allows for a more precise and clear reconstruction of the sources that generated the magnetic field as the tissues above the cerebral cortex (skull, scalp) are practically transparent to the magnetic field while they distort the field and electrical potentials measured by EEG.

Thanks to appropriate algorithms based on network theory, it is possible to use MEG to study the *cerebral connectivity* functional, which plays a fundamental role in understanding how the brain works. Although there is a clear specialization among the regions of the cerebral cortex, the true power of the brain seems to derive from the ability of those regions to work together on a

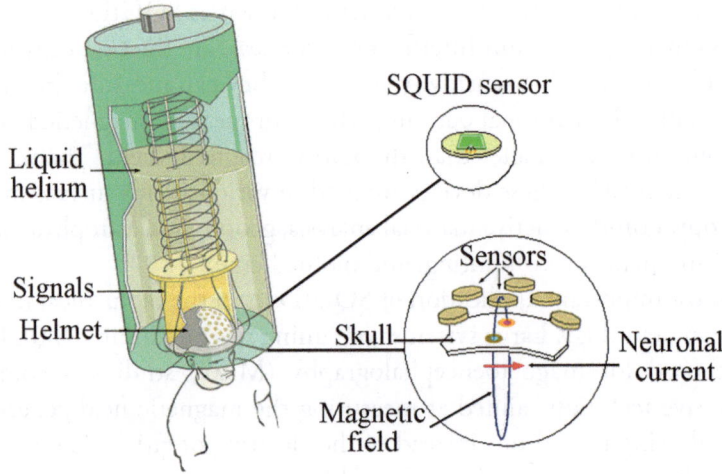

Fig. 3.11 Diagram of a system for Magnetoencephalography. The powerful SQUID sensors measure the magnetic field generated by the weak neuronal currents having an intensity of 10–100 fT (equal to a few billionths of the Earth's magnetic field). To cool the sensors below the critical temperature, liquid helium is used (T = 4 K = −269 °C)

range of spatial scales like a set of richly interconnected and complex networks. The aforementioned brain networks are made up of spatially distributed regions but functionally connected and process information, facilitating the integration and coordination of various functions.

Abnormal connectivity is linked to the presence of diseases such as neurodegenerative ones (for example Alzheimer's syndrome), which makes the study of connectivity of great interest also in the clinical field. Despite the considerable potential, MEG systems are mainly used for clinical research and basic neurological research.

Other applications of the Josephson effect involve elementary particle detectors, quantum metrology for the definition of the electrical voltage standard and the realization of quantum bits for quantum computation, which we will discuss in the next chapter.

Another noteworthy application of quantum physics is tunneling microscopy, but before discovering its charm, it is necessary to open a small parenthesis on microscopy in general. The invention and use of microscopes has been of fundamental importance in almost all scientific sectors allowing a significant step forward in the knowledge of natural phenomena. Thanks to microscopy, scientists have been able to confirm and verify their hypotheses and theories, looking with their own eyes at what they had only imagined. To observe ever smaller structures, sophisticated instruments based on different operating principles must be used. Optical microscopes allow observing structures up to a few thousandths of a millimeter. This is because, the wavelength of light being between 0.4 and 0.7 thousandths of a millimeter, when the objects to be observed are comparable or smaller than the wavelength of the light that is reflected from the sample, interference or diffraction phenomena typical of the wave nature of light appear, preventing clear and sharp images from being obtained. The electron microscope, also known as SEM (acronym for Scanning Electron Microscope), uses a beam of electrons whose speed is such as to obtain for the electrons a De Broglie wavelength much smaller than that of light and this allows observing objects up to a few millionths of a millimeter, therefore three orders of magnitude smaller than those observable with the optical microscope. The electrons reflected from the material are detected by suitable detectors, generating electrical signals that produce the image on a screen or computer monitor. In the case of samples with thicknesses of the order of 50–100 nm (less than one ten-thousandth of a millimeter), the electron microscope can also be used to observe the section of the sample. In this case, we speak of TEM (acronym for Transmission Electron Microscope) in which the electron beam completely crosses the material and hits a fluorescent screen in which the images of the section are projected with a resolution of about one Angstrom (0.1 nm), that is, a few ten-millionths of a millimeter. Such sensitivity allows even single atoms in solids to be observed.

The cryogenic version of the TEM (cryo-TEM), in which the samples are cooled to the temperature of liquid nitrogen (−196 °C), is very useful for studying macromolecules of biological and chemical interest. The cooling, in addition to limiting the damage of the electron beam, allows to preserve the hydration of the molecules so that it does not get the original shape and structure are compromised. This is easily understandable if we think, for example, about dried and frozen porcini mushrooms. In the first case, it is clear to us that the shape and structure of the mushroom is altered by the drying process, which essentially involves total dehydration; if instead the mushroom is frozen, the shape remains practically unchanged as it retains the amount of water that allows it to have that particular shape. Beyond the constructive complexities of a scanning or transmission electron microscope, we can certainly affirm that the quantum phenomenon on which it is based and that allows us to look very deeply inside matter, is the wave nature of the electron.

But the microscope on which we want to focus is based on the tunnel effect and allows us to acquire extraordinary images of the surfaces of materials with atomic resolution, allowing to see the individual atoms on the surface of the observed material. The Scanning Tunneling Microscope (STM), indispensable for the study of the surfaces of solids, was invented by Gerd Binnig and

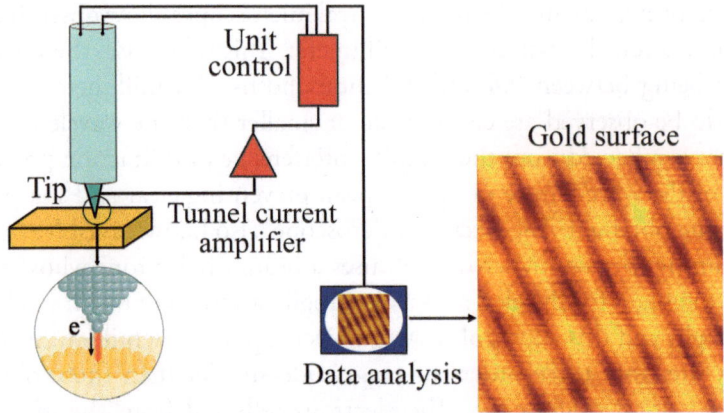

Fig. 3.12 Diagram of a tunnel effect microscope (STM). An extremely thin tip is brought very close to the surface of the material to be analyzed. Electrons migrate from the tip to the material due to the tunnel effect, allowing through specific algorithms to reconstruct the morphology of the surface. On the right is an image of a gold surface in which the individual gold atoms are clearly visible in light yellow

Heinrich Rohrer in 1981, who for this invention were awarded the Nobel Prize in Physics in 1986.

The microscope consists of a metallic tip that can be brought very close to the surface of the material to be examined (Fig. 3.12). By varying the distance between the metallic tip and the surface of the sample, it is possible to visualize the morphology of the surface structures.

If the distance between the tip and the material is sufficiently small, an electron can jump from one side to the other due to the tunnel effect; in fact, the two atoms of the tip and the material, although not directly touching each other, come to be slightly in contact at the corrugated region constituted by the electron cloud. There is therefore an overlap of the wave functions associated with the outermost electrons of the atoms and this determines, as seen in Chap. 2, the tunnel effect of the electrons. The intensity of the tunnel current, extremely sensitive to the distance between the microscope tip and the surface atoms, is detected during the scanning of the surface. In this way, the tip manages to follow the profile of a row of atoms, reconstructing the topography of the sample surface. With an STM, very high spatial resolution can be achieved up to 0.1 nm, just double the size of a hydrogen atom.

We conclude this brief review on the major applications of quantum physics with some mention of a nanotechnology topic which, in addition to being interesting from a physical point of view, is having a noticeable technological impact also on widely distributed products known as *quantum dots*, a topic for which the Nobel Prize in Chemistry was awarded in 2023 to Moungi Bawendi, Louis Brus and Alexei I. Ekimov.

These are semiconductor nanocrystals with a size of the order of a few billionths of a meter (1–10 nm). Having such small dimensions, comparable to the De Broglie wavelength of the electrons contained in them, quantum dots show marked quantum behaviors, the most evident of which is the presence of discrete energy levels, unlike macroscopic semiconductors in which, as seen in Sect. 3.1, energy bands exist. Therefore, even though they have the structure of a solid, quantum dots exhibit behaviors similar to individual atoms, for this reason they are considered *artificial atoms*. This behavior should not surprise us, as seen in the second chapter, the quantization of energy levels arises when a particle is forced to be confined in a space comparable to its De Broglie wavelength, like the electron in a hydrogen atom or as in this case in a *nano-box* constituted by the semiconductor single crystal.

In particular, when a quantum dot is excited with an electromagnetic radiation, for example ultraviolet light, the electrons move from the valence band to the conduction band forming electron-hole pairs. Due to the nanoscopic size of the semiconductor crystal, systems similar to individual atoms are formed in which the positive nucleus is constituted by the hole. These pseudo-atomic systems, known as *excitons*, present discrete energy levels whose separation is typically lower than the energy of the ultraviolet rays used to excite the quantum dot. Therefore, if the quantum dots are hit by ultraviolet light, excitons are formed which, decaying into lower energy levels, return a light with a lower frequency, typically in the visible spectrum (phenomenon of *fluorescence*, Fig. 3.13).

Another peculiar characteristic is given by the fact that the dependence of the separation between the fundamental level and the first excited level depends on the size of the quantum dot: as their size decreases, the distance between the two energy levels increases. For example, in a typical cadmium selenide (CdSe) quantum dot, this separation varies from 1.8 eV for the larger ones to 3 eV for the smallest ones. This energy range covers the entire optical spectrum of electromagnetic radiations, which is equivalent to saying that quantum dots can absorb and/or emit light in the entire visible spectrum, and by changing their dimensions we can choose the frequency at which the quantum dots emit and/or absorb light.

Since the separation between the energy levels is greater in small quantum dots (2–3 nm), the light emitted by these has a higher frequency (green, blue, purple), while those with larger size (5–7 nm) emit light with lower frequencies (red, orange, yellow). It should finally be said that the efficiency with which these nanomaterials absorb and emit light is extremely high, in

Fig. 3.13 Vials containing quantum dots of different sizes irradiated with ultraviolet light. The frequency of the emitted light increases as the size of the dots decreases

particular configurations they allow to emit almost all the absorbed light, resulting in particularly bright objects (Fig. 3.13).

Currently, quantum dots are used as high-quality light emitters in TV screens (QLED—Quantum dot led) and for biomedical imaging in which, thanks to their remarkable fluorescence, very intense and vivid images of cellular organelles or proteins are obtained. But many other applications such as high efficiency solar cells (over 60%), quantum computing and pharmacology, could be available soon.

Following this brief compendium on the applications of quantum physics, it is natural to be amazed by the fact that at the base of these precious technologies that have changed our way of living, there is the bizarre and strange quantum world, where particles are even waves have the gift of ubiquity, they appear and disappear, they transform into energy, they jump from one orbit to another and pass through barriers even without having the energy to do so. In short, a world apparently worthy of the best science fiction films!

But the quantum wonders do not end here. As we will see in the next chapter, they are driving another quantum revolution that will have a truly significant technological impact.

4

The Second Quantum Revolution and Quantum Technologies

On October 4, 2022, the Royal Swedish Academy of Sciences awarded the Nobel Prize in Physics to Alain Aspect, John F. Clauser, and Anton Zeilinger for their experiments on a distinctly quantum phenomenon known as quantum entanglement and for laying the experimental foundations of quantum information and computation.

In this fourth chapter, we will try to illustrate the fundamental concepts of this interesting and in some respects mysterious topic that has had and will have a great impact on current and future technologies.

4.1 The Copenhagen Interpretation and the Paradoxes of Schrödinger's Cat and Zeno's Quantum Arrow

The conceptual foundations of quantum mechanics have been and in some respects still are today, the subject of lively disputes and controversies due to different interpretations. When Erwin Schrödinger wrote his famous equation, he was undoubtedly inspired by the wave nature of matter proposed by De Broglie, so it seemed natural to him to interpret the wave function (solution of Schrödinger's equation) as an electronic wave since he initially wrote it for the electron. But the double-slit interference experiments (Fig. 2.3) tell us that the electron is detected on the screen as a particle, that is, when one or more electrons pass, we see one or more points on the screen and not a wave. The interference pattern begins to emerge when several hundred or thousands

© The Author(s), under exclusive license to Springer Nature Switzerland AG 2025
C. Granata, *A Journey into Modern Physics*, https://doi.org/10.1007/978-3-031-77775-2_4

of electrons have passed through the double slit. It is therefore evident that the interpretation of Schrödinger's wave function was contradicted by experimental facts. As already mentioned in the second chapter, the probabilistic interpretation of the wave function was soon proposed and accepted by most physicists, according to which the square modulus of the wave function represents the probability of finding the particle at a given point at a certain instant. The interpretation of the wave function is however only one aspect of a more general interpretation of quantum mechanics that also considers other aspects and which we will deal with in this paragraph.

The most widespread interpretation of quantum mechanics is that of Copenhagen or orthodox developed at the end of the twenties of the last century by most of the founding fathers of the new theory: Bohr (primarily), Born, Pauli, Heisenberg. In addition to the statistical-probabilistic meaning of the wave function, this interpretation assumes that in general a quantum system or state can be described by a wave function solution of Schrödinger's equation and that this function can be written from the superposition or linear combination of other wave functions that represent other quantum states each weighted with a certain probability (*principle of superposition*). A linear combination of a set of elements is nothing more than their sum in which each element is multiplied by an arbitrary number. If, for example, we consider two solutions Ψ_1 and Ψ_2, even the infinite functions given by $a\,\Psi_1 + b\,\Psi_2$ with a and b arbitrary numbers are still solutions of the Schrödinger equation. After all, from a mathematical point of view the linear combination of two or more solutions of the Schrödinger equation is still a solution of the equation. In this sense, the principle of superposition is a mathematical consequence of the fundamental equation of quantum mechanics. In some cases the solutions are infinite and therefore any linear combination of these infinite solutions is still a solution. Starting therefore from a set of solutions, you can construct infinite functions still solutions of the Schrödinger equation, just as the pieces of a lego can be assembled to construct many structures. These wave functions, linear combinations of various wave functions must however satisfy the normalization condition (see Chap. 2), which implies that the sum of the squares of the numerical coefficients must be equal to 1. Therefore the square of each coefficient can assume a value between 0 and 1. The closer a coefficient is approaches 1, the greater will be the weight of the relative wave function associated with it. The mathematical functions solutions of the Schrödinger equation can be considered as vectors in a space with n dimensions where n can also be infinite. We remember that in the most common sense, a vector is a segment oriented in three-dimensional Euclidean space endowed with intensity proportional to its length, direction and sense. Any vector in space, can be written as the linear combination of three unit length vectors and perpendicular to each other, that is three vectors that lie along the three Cartesian axes. In the same way any wave function or state vector, solution of the Schrödinger equation, can be written as a linear combination of vectors perpendicular to each other and in turn solutions of the Schrödinger equation. For example, if we consider the hydrogen atom, there are infinite solutions of the Schrödinger equation and each of them is linked to one of the infinite energy levels of the electron. So in principle, a linear combination of the infinite solutions is still a possible state for the hydrogen atom.

The crucial point is the interpretation attributed to the principle of superposition and measurement. According to the Copenhagen interpretation, if a system is described by a wave function that is the superposition of two or more states, it is not possible in any way know, before making a measurement, in which state the system is located. At the moment of measurement the wave function collapses into one of the states of which it is composed; before the measurement it makes no sense to ask in which state the system is located, at most it can be said that it is simultaneously in all states but with different probability depending on the state (*quantum coherence*). According to the orthodox interpretation, the superposition of quantum states is an intrinsic and irreducible property of nature not due to the limited knowledge of the physical conditions of the system under consideration.

You can imagine the superposition of two quantum states as something analogous to what is observed when looking at some paintings in which two different images coexist, such as for example the vase of Edgar Rubin (Danish psychologist and philosopher, 1886–1951), shown in Fig. 4.1. If you look at the painting from right to left and vice versa you observe two profile faces, if instead you look at it from bottom to top and vice versa, we notice the presence of a vase.

Another example of a bistable image is represented by the duck-rabbit (Fig. 4.1), proposed in 1892 by the American psychologist Joseph Jastrow (1863–1944) to demonstrate an optical illusion. Also in this case, the figure is composed of a single image that, alternatively, can be interpreted as a rabbit looking to the right or a duck looking to the left. Obviously, this is just an analogy, also because after observation the images continue to coexist, instead in the case of a measurement on a two-state quantum system, only one survives.

In the particular case of the position of a particle described by a wave function, it makes no sense to ask in which point of space the particle is before the measurement or it is simultaneously in all positions and the measurement process collapses the wave function at the point where the particle is found after the measurement (Fig. 4.2).

Fig. 4.1 Examples of two bistable images in which two different images coexist. On the left, the Rubin vase represents simultaneously two profile faces and two vases, while on the right the duck-rabbit represents a rabbit looking to the right and/or a duck looking to the left. In quantum physics, something similar happens with two quantum states, before the measurement they are simultaneously in both states

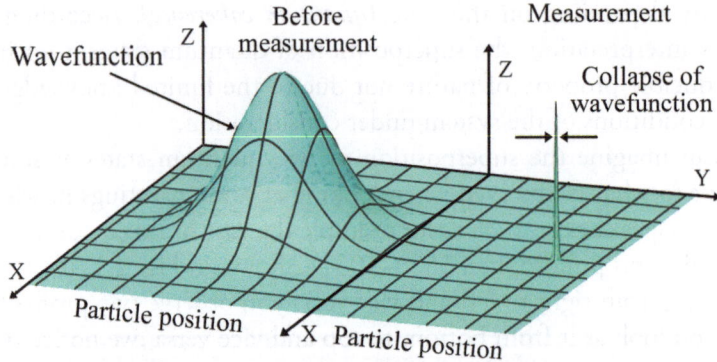

Fig. 4.2 Collapse of the wave function: before the measurement the function is delocalized in a wide region of space, after the measurement it becomes very peaked and localized at the point where the particle is detected

From what has been said, it can be deduced that according to the Copenhagen interpretation, the role of measurement is fundamental to discover the characteristics of a system and the measurement instrument becomes one with the system to be measured.

To underline the absurdity of certain questions in quantum mechanics, the physicist and philosopher of science David Albert argued that asking where the particle is before measuring it is equivalent to asking if an elementary particle is married! As we will see shortly, when we consider two particles, to the gift of ubiquity of quantum particles are added other even more extraordinary properties that lead us to think that perhaps it makes sense to ask if a particle is married or more correctly if it is indissolubly bound to another particle.

Returning to the collapse of the wave function, it should be clarified that it applies to any measurable physical quantity such as energy, angular momentum, spin, etc. A quantum system can be in a superposition of states with different energy, angular momentum, or spin and only at the moment of measurement will we know the value of the physical quantity we are considering. Before the measurement we can only say that we have a certain probability of finding a certain quantity. This probability is equal to the square of the coefficient related to the wave function corresponding to that quantity. Continuing to consider the example of the hydrogen atom, if it is in a generic state, linear combination of the infinite states related to the quantized energies of the electron, before the measurement it is not possible to say what energy the atom has, the measurement operation forces the atom to assume one of the available energies i.e. those corresponding to a certain principal quantum number n. Any other measurements always give the same result, unless the hydrogen atom is returned to a generic state. In practice, a hydrogen atom, at room temperature, is always in the level fundamental and whatever measure of energy is taken, the same energy will always be found. This is because thermal energy is much smaller than the energy needed to excite the electron to the next orbital.

It is worth emphasizing once again that quantum mechanics is not a statistical/stochastic theory but is extremely deterministic, in the sense that we can know the exact spatial and temporal evolution of the wave function by solving the Schrödinger equation. However, the measurement process, as previously mentioned, in the case where there is an overlap of states, includes an intrinsic uncertainty linked to the mysterious collapse of the wave function into one of the states it is composed of.

But not always are quantum systems in an overlap of states, for example, as mentioned above, the energy states of the electrons of an atom, at room temperature, are not in an overlap of various states and if a measurement of the energy levels is made, the same values are always found which are in perfect agreement with theoretical predictions, that is, the energy of each electron is known exactly even before making the measurement. But even when using the probabilistic aspect of the wave function or quantum statistics (see Chap. 1), predictions can be made with astonishing precision. Consider that in the field of quantum electrodynamics, there are quantities like the anomalous magnetic moment of the electron whose calculated value has a precision of one part in a billion, that is, if we compare the calculated value and the measured one, the first deviation we observe is after 9 decimal places!

In summary, a quantum system can be in an overlap of states and nothing can be said before making a measurement, at most one can estimate what is the probability of finding a certain state after of the measurement. The measurement process destroys the overlap of states forcing the system to collapse into one of the possible states. From this, the fundamental role of measurement in quantum physics can be inferred, as well as its intimate connection with the physical process being observed. After all, as mentioned in the second chapter about the double-slit experiment, if you even non-invasively watch the electron as it passes through the double slit, the wave aspect of the particle disappears and only the corpuscular one remains.

Einstein's position was strongly opposed to the Copenhagen interpretation. In particular, he argued that there were hidden variables (*hidden variable theory*) that could not be determined and that were the origin of the statistical-probabilistic behavior of quantum phenomena, therefore, according to the German genius, quantum mechanics was an incomplete theory. Einstein's position is well described by the famous phrase "God does not play dice" or by the ironic question he asked Bohr, a staunch supporter of the Copenhagen interpretation: "Do you really believe that the moon exists only if you look at it?" Einstein was deeply convinced that the laws of nature should be local, that is, every phenomenon, body or more generally a system must be conditioned only by what happens in the immediate vicinity. As already mentioned in the first chapter, Einstein was also very critical of Newton's universal gravitation theory which predicted a force at a distance. Einstein argued that if the Sun disappeared, the Earth would not immediately go out of orbit, but would do so after the time necessary for light to reach the Earth from the Sun, that is, about 9 min. And it is precisely this rooted position of local realism that will lead Einstein to take a very skeptical and critical attitude towards the conceptual foundations of quantum mechanics.

Schrödinger himself was not very convinced of the orthodox interpretation. In fact in 1935 he devised a thought experiment known as Schrödinger's cat paradox, in which he imagined enclosing a cat in a box containing a vial of cyanide that would break if one of the atoms present in a radioactive substance disintegrated, emitting radiation that triggered the vial-breaking mechanism (Fig. 4.3). Assuming that the probability of a radioactive atom disintegrating was 50% at a given time, Schrödinger argued that according to the orthodox interpretation of quantum mechanics, before opening the box the cat was in a superposition of the two states, both alive and dead, and when the box was opened, the wave function describing the cat collapsed into one of the two possible states: alive or dead. It is clear that such a superposition of alive/dead states is never observed, hence the paradoxical aspect of the

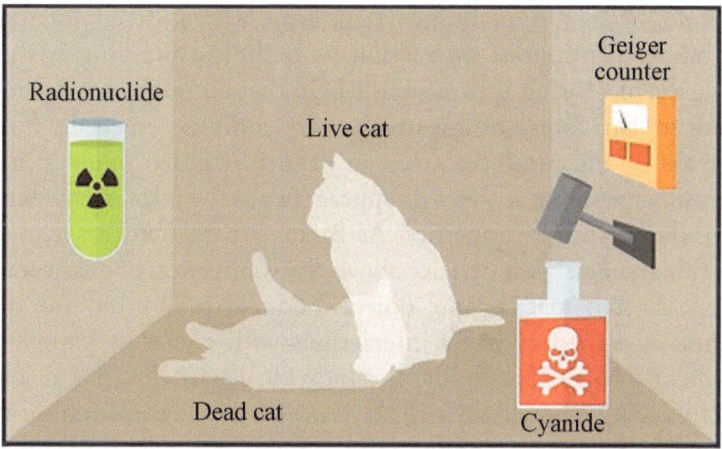

Fig. 4.3 Representation of Schrödinger's cat paradox. According to the Copenhagen interpretation, the cat is simultaneously alive and dead and only when we open the box we induce one of the two states (alive or dead) through the collapse of the wave function

experiment. The paradox highlighted two fundamental conceptual aspects. Does quantum mechanics apply at the macroscopic level or can we only use it to describe microscopic phenomena? If so, how do we explain phenomena in which there is interaction between the microscopic world (disintegrating atom) and the macroscopic world (breaking vial)?

Many experiments conducted between the end of the last century and the beginning of this century have unequivocally shown that quantum mechanics also applies at the macroscopic level, clearly demonstrating some macroscopic quantum effects such as the quantum superposition of different macroscopic states through the measurement of the coexistence of a small current in a superconducting ring in two distinct states: clockwise and counterclockwise current. An experiment that leaves little doubt about the existence of macroscopic quantum effects was carried out in 2023: the superposition of two states was observed in a crystal of piezoelectric material weighing 16 μg, corresponding to about 100 trillion atoms, a decidedly macroscopic object (a 16 μg Schrödinger's cat!). A piezoelectric material is able to deform/oscillate if an electric field is applied to its ends. The piezoelectric crystal in question, subjected to an appropriate electric field also in a superposition of states, was made to oscillate simultaneously in two opposite directions, reproducing Schrödinger's thought experiment.

So how is the paradox resolved in the light of the Copenhagen interpretation? Currently, the most convincing solution is provided by the theory of *quantum decoherence*. A quantum system must be considered not an isolated system but in continuous interaction with the surrounding environment including the measuring apparatus, the light that illuminates it or the air that surrounds it. This interaction is usually very complex, especially for macroscopic systems, and produces a sort of disturbance that tends to make the coherent superposition of states disappear (hence the term decoherence) and therefore the quantum properties. As Brian Greene states in his book *The Fabric of the Cosmos*: "decoherence allows the strangeness of quantum physics to disappear from macroscopic objects because, bit by bit, the quantum strangeness is carried away by interactions with countless particles of the environment". More specifically, the individual particles of a macroscopic system, following interaction with the particles of the surrounding environment, modify their wave function independently of each other and the delicate superposition of states quickly disappears, losing the necessary coherence that is at the basis of the quantum behavior of a set of particles. The greater the number of particles that make up the quantum system, the faster the process of decoherence. In relatively simple quantum systems that are accurately isolated from the surrounding environment, we talk about record times of the order of a thousandth of a second. In the case of an extremely complex macroscopic system like the quantum cat made up of thousands of billions of billions of atoms, the decoherence time would be so small as to be impossible to measure. Moreover, the same interaction between the various molecules that compose it produces decoherence and cancels out quantum phenomena. Decoherence is the reason why almost all the macroscopic world appears to us as classical, and not quantum, that is, populated by Schrödinger's cats!

Another paradox related to the problem of measurement in quantum mechanics is that of Zeno's arrow or also known as *Zeno's quantum effect*. Zeno of Elea, the same one from the famous paradox of Achilles and the tortoise, devised another paradox, that of the arrow. He imagined shooting an arrow at a target and observing the arrow at a certain moment, arguing that, if a precise moment is fixed, the arrow appears still like the frames of a film. Since time is made up of many moments one after the other in which the arrow is stationary, Zeno deduced that the arrow was stationary and could never reach the target. Like the paradox of Achilles and the tortoise, the paradox is easily resolved with infinitesimal calculus, pointing out to Zeno that what he considered moments are actually very small time intervals (infinitesimal), and in these infinitesimal intervals the arrow covers an infinitesimal stretch. Since the sum of infinite infinitesimal intervals, according to differential calculus, gives a finite number, the arrow reaches the target in a finite time.

From a quantum point of view, this paradox has inspired the following effect/paradox: if we consider a system described by a wave function at a certain moment, a superposition of various states, quantum mechanics tells us that the evolution of the temporal wave function is described by Schrödinger's equation but also tells us that the moment a measurement is made the system collapses into one of the possible states that make up the wave function, so the measurement somehow prevents the evolution of the wave function. The same mechanisms of radioactive decay are linked to the evolution of a quantum state, which can be inhibited if repeated and very fast measurements are made before the atom decays. This implies that, in principle, it is possible to slow down a decay process or even block it. Therefore, just as Zeno's arrow stops when it is observed at a given moment, some processes can be slowed down or blocked if their state is disturbed by fast and repeated measurements before it evolves. Although the phenomenon is indeed paradoxical, experiments have been carried out that seem to show the slowing down/freezing of some quantum processes and therefore the existence of the Zeno effect. Among the most important experiments in this sense, we recall that of 1990 on Be 9 atoms in which a freezing of the atom in an energy level was observed, activating an appropriate laser source that blocked the decay.

Two other interesting experiments on the Zeno effect were carried out in 2014 and 2015 on a Bose-Einstein condensate of rubidium atoms. In the first experiment the freezing of the dynamics of the atoms was observed through a strong disturbance, while in the other the formation of zero spin particles resulting from the union of two particles with equal and opposite spin was inhibited by interaction with an appropriate laser. We can therefore conclude that, with a good probability, if poor Schrödinger's cat knew how to make repeated and very fast measurements on the radioactive source before it decays breaking the cyanide vial, it would always come out of the box alive!

4.2 EPR Paradox, "Entanglement" and Alternative Interpretations of Quantum Mechanics

Let's get to one of the most controversial and bizarre phenomena of quantum mechanics, the *quantum entanglement* or *quantum correlation*, a term introduced by Schrödinger. The inventor of the famous fundamental equation of quantum mechanics considered entanglement the most peculiar phenomenon of quantum physics.

In this regard, we remember another famous paradox devised by Albert Einstein, Boris Podolsky, and Nathan Rosen (also in 1935) based on an ideal experiment and known as the *EPR paradox* from the initials of the three authors. We will consider the simplified version formulated by the American physicist and philosopher David Bohm.

Imagine that from a source in a box, at a certain instant, two particles with spin are emitted in the same direction but with opposite signs (Fig. 4.4).

According to the law of conservation of total angular momentum, the total spin (see Chap. 1) of the system consisting of the two particles must be conserved and in this case must be zero as it is zero before the emission of the particles. Therefore, one particle must have spin up and the other spin down. According to quantum mechanics, the two particles (EPR pairs) can be described by a wave function that is the superposition of two states: one in which the particle traveling to the right has spin up and the one traveling to the left has spin down, the other state is the inverted one, i.e., the right particle with spin down and the left one with spin up, or the one represented in Fig. 4.4.

The Copenhagen interpretation tells us that it makes no sense to ask in which state the two particles are before the measurement and it is the act of measurement that collapses the wave function into one of the two states. This means that, even if the particles are thousands of kilometers apart, the moment the spin on one particle is measured, the wave function collapses and the other particle instantly assumes the opposite spin. There is therefore a kind of telepathic bond between two quantum correlated (entangled) particles, what Einstein, in complete disagreement with this interpretation, called "spooky

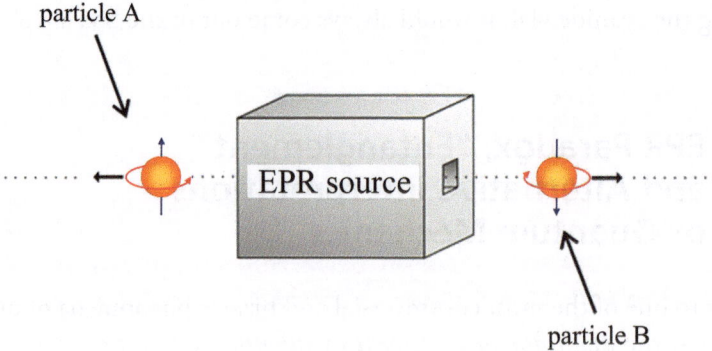

Fig. 4.4 Scheme of the EPR paradox: the particles are emitted in the same direction but with opposite signs. The measurement of the spin state of one particle instantly determines the state of the other particle (entanglement)

action at a distance". Moreover, according to Einstein, this implied a violation of the principles of Relativity and in particular the impossibility of traveling at a speed greater than that of light.

Therefore, according to Bohr and his followers, quantum mechanics is a non-local theory in the sense that what happens at one point instantly conditions what happens at another point even at great distances, while according to Einstein this is not possible and behind the telepathy of the particles or the quantum indeterminacy there is simply an incomplete knowledge of all the variables of the system.

To the famous EPR paradox, Bohr replied that there is no information that propagates at a speed greater than light since the spin of one of the particles is determined only when the measurement is made. In particular, if the particle traveling to the right is measured by an observer A resulting in spin up, until observer B measures the particle traveling to the left will never know in which spin state the other particle is. So it must wait for the particle traveling at a speed lower than that of light to arrive, or wait for the observer A to send a message telling it that the left particle has a spin down since its own has a spin up.

Well, at this point it is necessary to make a clarification so as not to confuse quantum correlation with classical correlation. If we imagine drawing two marbles of different colors (red and blue) from a closed-eye urn and throwing them in opposite directions, one to the right and the other to the left, we do not know before opening our eyes in which direction we have thrown the red or blue marble. Suppose we turn to the right and open our eyes, we then discover the color of the marble that was thrown to the right and consequently even if we do not turn to the left we know with certainty the color of the other marble. This is also an example of correlation between two bodies, but in this case it is not an *entangled* system, anyone observing our trivial experiment observes the direction of the marbles from the beginning and before we open our eyes knows the direction of the two colored marbles. In reality, already at the moment of the throw the blue marble was going in one direction and the red one in the other and this depends on how the urn was shaken, how and when the marbles were taken, and other factors impossible to evaluate exactly and that give the experiment a character of randomness. This is certainly the case of the hidden variables that Einstein talked about, but in the microscopic world or in particular macroscopic systems, quantum mechanics tells us that before making the measurement and thus inducing the collapse of the wave function it makes no sense to ask what type of particle went in one direction or another and the moment one of the two particles is detected, it instantly induces the other particle to assume one of the two states.

The concept of quantum correlation is difficult to digest, not because it is difficult to understand but because it is completely out of common sense and is in some respects a dogmatic concept. The dogma lies in the collapse of the wave function, which remains one of the unexplained mysteries of the foundations of quantum mechanics.

Speaking of the collapse of the wave function, it is certainly instructive to mention another paradox of quantum mechanics less known than that of the famous cat but equally interesting: the *Wigner's friend paradox*. The paradox concerns the problem of measurement in quantum mechanics as well as the apparent contradiction due to the deterministic evolution of the wave function on the one hand and the probabilistic nature of the measurement on the other. According to Wigner, if in an isolated laboratory there is a friend of his who is making measurements on a quantum system and he observes it from the outside, the isolated system formed by his friend and the experiment he is conducting is simultaneously deterministic as the system is isolated and probabilistic since Wigner's friend is making measurements. The paradox could easily be resolved by stating that the laboratory and Wigner's friend are macroscopic systems for which quantum mechanics does not apply, but we know well that many macroscopic systems show marked quantum properties.

The mathematician and philosopher John von Neumann in his 1932 treatise, *The Mathematical Principles of Quantum Mechanics*, proposes that the consciousness of the observer would be at the origin of the collapse of the wave function. This somewhat metaphysical hypothesis solved the paradox of Wigner's friend by simply asserting that the collapse is caused by the first conscious observer, i.e., the friend who is making measurements in the closed laboratory.

The other solution is to bring up decoherence as done for Schrödinger's cat or try to find a theory that can take into account both classical macroscopic systems and quantum microscopic ones, and provide an explanation for the collapse of the wave function. One of the most important theories that proposes this ambitious goal was the *GRW theory*, named after the initials of the three physicists who proposed it: Giancarlo Ghirardi, Alberto Rimini, and Tullio Weber. The main purpose of the theory is to objectively explain the mysterious collapse of the wave function, thus providing an alternative interpretation to the orthodox Copenhagen interpretation. The central idea of the GRW theory is the assumption that the process of wave function collapse is spontaneous and not linked to measurement, that is, a particle localizes spontaneously without any observer or measuring instrument stimulating the process as predicted by the Copenhagen school. The spontaneous collapse of the wave function, according to the GRW theory, depends on the ratio between the number of particles N and a very large time τ (on the order of 10^{15} s), therefore for macroscopic systems in which there are a number of particles comparable to Avogadro's number (6.022×10^{23}), the probability of spontaneous collapse is very high, the collapse is almost instantaneous. Instead, in the case of a single particle or a set of particles not excessively large, the probability of spontaneous collapse remains very small. Naturally, experiments in macroscopic quantum mechanics, like those mentioned in the previous paragraph, have somehow challenged the validity of the theory.

The GRW theory is not the only alternative interpretation to the orthodox one, there are others and it may be worth opening a small parenthesis to say something about the other alternative interpretations.

The Copenhagen interpretation remained and remains the most widespread, but, in addition to the aforementioned position of Einstein and the GRW theory just described, over the years several other interpretations were born, some even by some of the founding fathers of quantum mechanics, who showed a certain intolerance to what they considered the "Copenhagen dogma". Among the most interesting is the De Broglie-Bohm interpretation, initially developed by De Broglie at the same time as the Copenhagen one took hold. De Broglie's hypothesis, presented at the fifth Solvay conference (Brussels, 1927), consisted in considering the wave function not as a purely mathematical entity, but a kind of real wave that determines the motion of particles on defined trajectories. The so-called "pilot wave" of De Broglie, indeed piloted the particles in a causal and deterministic way. Unlike the wave-particle dualism for matter, this new idea of De Broglie was practically ignored. About 25 years later, David Bohm took up the idea of the French scientist, deepened it and extended it, showing that it was possible to take into account all the experimental results with this alternative interpretation. In fact, both the Schrödinger equation and the entire mathematical structure remained the same, but the conceptual assumptions changed. For Bohm, the wave function equation contained information related to a "quantum potential", a kind of energy that propagates in space and determines the motion of particles on certain trajectories, explaining the apparent strangeness shown by quantum particles. In the case of the double-slit experiment, which we talked about in the first chapter, the electron passes through a single slit guided by the quantum potential, which interacting with the double slit sets up an interferometric path for the electrons. The moment we try to see which way the electrons are going, we disturb the quantum potential and the interference disappears. Moreover, any disturbance of the quantum potential is instantly transferred throughout space, giving Bohm's theory a strong aspect of non-locality.

In Bohm's theory, particles are well defined and localized and no longer extended and delocalized, and the dogmatic assumption of the collapse of the wave function falls, as well as the nonsense of asking where the particle is before measuring it, axioms at the base of the Copenhagen interpretation. Moreover, the impossibility of accurately determining the trajectory of a particle is linked to the lack of knowledge of some variables of motion (Einstein's hidden variable theory). Therefore, like Einstein, according to Bohm, quantum mechanics was an incomplete theory that prevented a deterministic description of nature, but unlike Einstein, Bohm preserved the non-local aspect of the theory.

There are further interpretations of quantum mechanics, somewhat more metaphysical in some respects, whose aim is always to propose an alternative vision to that of Copenhagen. Among these, it is worth mentioning the one proposed in 1957 by the American physicist Hugh Everett III, also known as the *many worlds* interpretation. As the name suggests, Everett's extravagant idea involves the existence of infinite parallel universes, each of which corresponds to a possible quantum state. Therefore, the principle of superposition is nothing more than the showcase in which all possible parallel universes are displayed, overlapping each other. Returning again to the double-slit experiment, in one universe the particle passes through one slit and in a second universe it passes through the other. There are therefore two possible universes for the electron that overlap in the region of the double slit where interference occurs. Or in the case of the entanglement just described, in one universe the particle going to the right has spin up and the other going to the left has spin down, and in the other parallel universe the complementary situation occurs. Due to decoherence, the two universes separate, no longer interact with each other and we observe only one of the two possible universes. Therefore, the mystery of the collapse of the wave function and of the measure is solved, replacing it with that of parallel universes, perhaps no less shocking and surreal. This interpretation has not been very successful, nevertheless some of its variants have followed, such as the *multiverse* interpretation, the *many histories* interpretation and the *many minds* interpretation.

We can therefore affirm that the peculiar and extravagant aspects of quantum physics have given rise to many debates and many interpretations of quantum mechanics of a philosophical and metaphysical nature, but according to the more pragmatic wing of quantum mechanics they had to be circumscribed and not confused with its extraordinary predictive capacity. In the world of quantum physicists, the warning, perhaps erroneously attributed to Feynman, spread: "Shut up and calculate". In other words, forget about the philosophical aspect and focus on the pragmatic one inherited from the Galilean experimental method.

4.3 Bell's Inequality and Aspect's Experiments

A turning point came in 1964 when the Irish physicist John Bell found a rigorous way to understand who between Einstein and Bohr was right, bringing the discussion on the conceptual foundations of quantum mechanics from a philosophical/metaphysical level to a much more pragmatic one of experimental physics.

Bell was very fascinated by Bohm's theory which finally managed to explain quantum phenomena in terms of particles that moved following precise orbits as happened in classical physics and did so regardless of the observer. However, while solving the problem of objective realism raised by Einstein, Bohm predicted even more markedly non-locality, so hated by Einstein who defined it, as already mentioned, a spooky action at a distance. Bell tried to eliminate non-locality from Bohm's mechanics but his attempts were in vain and so he had the serious suspicion that this was not possible and that quantum mechanics was intrinsically non-local. He then began to work to find a rigorous way to demonstrate that it is not possible to construct a quantum theory eliminating non-locality.

Specifically, in a famous 1964 article titled *On the Einstein Podolsky Rosen Paradox (Einstein, Podolsky, and Rosen paradox)*, he formulated mathematical inequalities known as *Bell's inequalities* that, if violated, would have unquestionably proven Bohr right and therefore the vision of quantum mechanics that predicts this sort of telepathy at a distance between particles, if instead they were preserved, the reason would have been Einstein's and therefore the interpretation of the hidden variables. Without the use of a minimum of mathematical formalism, it is not easy to explain what these inequalities consist of, and there is a serious risk of making verbose and tangled speeches that most likely would only result in confusing the reader. In any case, without going into these details, Bell starts from the EPR thought experiment done with photons characterized by different quantum states (*polarization*) and constructs a *correlation function* whose value takes into account how many times two distant photons will be measured in the same state. The aforementioned function assumes a value less than *1* if there is no quantum correlation, i.e., if the theory is local as Einstein argued. If, on the other hand, non-local effects related to entanglement are considered, a value greater than *1* is found, violating the aforementioned inequality. Equivalently, we can say that, in the case of violation of the inequality, any local hidden variable theory is incompatible with quantum mechanics and the telepathy between the particles *entangled* is real. However, Bell's theory does not prohibit that quantum mechanics may have an interpretation in terms of hidden variable theory, but it tells us that whatever its interpretation must necessarily be non-local.

Bell had therefore thrown down a challenge to experimental physicists to carry out experiments that could confirm whether quantum mechanics violated or not his inequality. Such experiments were not easy to carry out and it took over 15 years for technological progress to allow for repeatable and reliable experiments.

In 1981 and 1982, Alain Aspect (Nobel laureate in Physics in 2022) and his research group carried out three important experiments in which they verified the violation of Bell's inequalities using, as a pair of particles, two photons *entangled*. As mentioned above, photons are the quanta of light and as such obey the laws of quantum mechanics, and for this reason it is possible to construct pairs of photons *entangled* using the state of polarization, i.e., the direction in which the electric field oscillates. Since the concept of light or photon polarization is fundamental also for the continuation of the chapter, we open a small parenthesis to provide some elements related to this characteristic of light.

In the first chapter we saw that light is an electromagnetic radiation with a certain range of wavelengths (0.4–0.7 μm) and we also know that an electromagnetic wave consists of a pair of electric and magnetic fields perpendicular to each other that oscillate at a certain frequency given by the ratio between the speed of light and the wavelength. The two electric and magnetic fields in addition to being orthogonal to each other are also orthogonal to the direction of propagation (transverse waves). Typically in the light that comes from the Sun but also in that of a light bulb, the electric and magnetic fields oscillate in the entire plane perpendicular to the direction of light propagation and not along an axis (for example the horizontal axis, the vertical one or an oblique one). In this case we speak of non-polarized light but if you use special optical filters it is possible to obtain an electromagnetic wave whose oscillations of the electric and magnetic fields occur only along one direction (Fig. 4.5).

By convention, we always refer to the direction in which the electric field oscillates. For example, if a vertical filter is used, at the output of the filter the electric field will only oscillate along the vertical axis and consequently the magnetic field along the horizontal axis. In fact, the filter cuts the oscillations of the electric field in all directions except those along the vertical axis. Obviously, filters can be used to obtain polarized light in all directions:

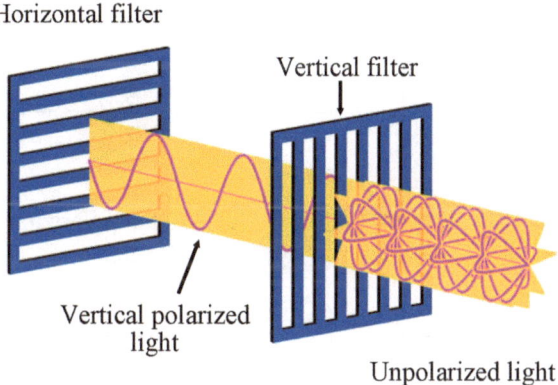

Horizontal filter

Vertical filter

Vertical polarized light

Unpolarized light

Fig. 4.5 Scheme of a polarizing filter. The non-polarized light passes through a vertical filter that eliminates all components of the electric field except those that oscillate in a vertical plane. Since the magnetic field is orthogonal to the electric one, it will oscillate in a horizontal plane

vertical, horizontal, oblique at a certain angle, or even simultaneously in two directions both vertical and horizontal or oblique with two different angles. Therefore, the use of filters allows us to select the directions in which we want the electric field to oscillate by eliminating the others. Polarizing filters are widely used in photography and in some types of glasses to block polarized light reflected from some surfaces such as glass and water mirrors, but they are also used in 3D cinema glasses, allowing us to see a three-dimensional image.

Returning to Aspect's experiments, in order to verify Bell's theory, a source of calcium atoms was used whose decay produced a pair of entangled photons *entangled* that moved along opposite paths and were detected at a distance of 13 m from each other (Fig. 4.6). Before being detected, the photons passed through polarizing filters that could both be vertical, horizontal, oblique or mixed, for example one vertical and the other horizontal as those represented in Fig. 4.6.

The polarization of the photons was for both vertical or horizontal to the direction of propagation of the photons, therefore in this state *entangled* the measurement of the polarization of a photon would have had to allow to deduce and predict exactly the polarization of the other. In practice, if the two filters had the same polarization, the entangled photons were both detected if they had the same polarization of the filters or none of the two was detected. The signals from the detectors were sent to a coincidence counter to evaluate the correlation of the pairs of photons. Analyzing the experimental data, it was deduced that the measurement of the state of one photon allowed to instantly read the state of the second photon as predicted by the orthodox

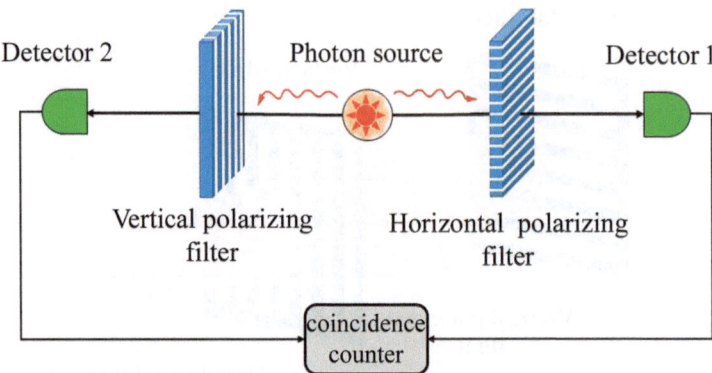

Fig. 4.6 Scheme of Alain Aspect's experiment to verify Bell's inequalities. Two photons traveling in opposite directions were made to pass through polarizing filters and subsequently detected. A coincidence counter checked the correlation between the pairs of photons. In the case shown in the figure, since the polarizers are oriented orthogonally to each other, the detection of a vertical photon by detector 2 did not correspond to an equal detection of the twin photon by detector 1, as the photons in the experiment had the same polarization (vertical or horizontal)

interpretation of quantum mechanics. Therefore, Bell's inequality was violated and it was experimentally demonstrated that quantum mechanics was an intrinsically non-local theory. Which is equivalent to saying that Einstein was wrong and the spectral action at a distance was real!

Starting from Aspect's pioneering experiments, there have been numerous experimental confirmations of the violation of Bell's inequality even at significantly greater distances. In particular, in 2017 using satellites, the phenomenon of entanglement was demonstrated on a pair of photons at a distance of 1203 km and in 2022 the mysterious effect was demonstrated on single atoms of rubidium at a distance of 33 km.

The experimental evidence therefore seems to support Bohr by clearly showing that quantum mechanics is a non-local theory in which what happens on a system is not necessarily caused by the immediate vicinity but can also be due to events that occurred thousands of km away.

In other words, the numerous experiments demonstrate that the mysterious telepathy between particles is real. Of course, in this case we can repeat what was said about Schrödinger's cat paradox: the phenomenon of decoherence does not allow us to observe in everyday life these "spectral actions at a distance"!

Some studies seem to indicate that quantum phenomena such as entanglement and quantum coherence are naturally used also by some living beings like the European robin to orient themselves during its long journey from

Scandinavian areas to Mediterranean ones as the cold season approaches, to the point of earning the nickname of *quantum robin*. These extraordinary birds are equipped with sensitive protein magnetoreceptors also known as *cryptochromes* present in the cells of the retina of the eye and that act as a magnetic compass to guide them along their long journey. The operation of this special compass could be based, according to these studies, on a subtle quantum effect linked to the coherent superposition of quantum states. When cryptochromes are hit by blue light, a photochemical reaction induces the jump of an electron from a biomolecule present inside them to another very close and always internal to the cryptochrome, thus creating a pair of charged molecules (radical pairs). Since both excited molecules have an odd number of electrons, the total spin of each molecule is not zero making possible two possible quantum states: one in which the spins of the two molecules have opposite directions and the other with both spins facing the same way. At this point a coherent oscillation between the two quantum states would be triggered, that is, a very fast quantum dance in which the pair of molecules switches from one state to the other in just under a millionth of a second. This quantum dance is influenced by the Earth's magnetic field that interacts with the spins of the molecules and when, due to the inevitable phenomena of decoherence, the process ends (after about 1 ten-thousandth of a second) and the biomolecules return to their initial state, the signals that reach the optical neurotransmitter depend on the bird's orientation in the Earth's magnetic field. This very fast process would be repeated continuously, allowing the robin to accurately identify its flight direction relative to the magnetic north and therefore to make any corrections.

At this point it is natural to ask whether the aforementioned experiments on *entanglement* have put a tombstone on debates related to the foundations of quantum mechanics. Bell's inequality and Aspect's experiments have certainly dissolved any doubt about the existence of non-local phenomena such as *entanglement* and the impossibility of explaining it in terms of a local theory with variables hidden. However, reflections on the conceptual foundations of quantum mechanics have continued and are still today the subject of interesting debates and interpretations.

In this regard, we recall the position of a well-known theoretical physicist of our time, Carlo Rovelli, who proposes a *relational interpretation* of quantum mechanics focusing on the importance of the interaction of an object or quantum system with the surrounding world: only interaction makes a quantum system real. In other words, quantum systems exist as they interact with other systems and there is no intrinsic or absolute quantum state. As he argues in his book Helgoland: "Objects are characterized by the way they interact…and there are no properties outside of interactions". Therefore, it makes no sense to talk about the characteristics of a completely isolated system, what we observe of a system is always the result of an interaction with some other system and this interaction changes from system to system defining its properties from time to time. About the infamous Schrödinger's cat, according to the Copenhagen interpretation, for us who observe the box from the outside, the cat is neither alive nor dead, that is, it is in a superposition of states. However, according to Rovelli's relational interpretation, the interaction defines the properties of the system changing depending on the objects or systems with which it interacts. The interacting system consisting of the cat, the radioactive atom and the cyanide bottle could have a definite outcome and therefore the cat could be definitely alive or dead regardless of external observers, that is the cat is alive or dead before the box is opened.

The possibility of being able to actively manipulate the quantum states of matter and in particular the *entangled* states has given rise to what is known as the *second quantum revolution* which is allowing the development of quantum technologies destined to have a significant impact on our society.

4.4   New Quantum Technologies

In this concluding paragraph, we will try to give a brief overview of the new quantum technologies also called *quantum technologies 2.0*. Of course, quantum technologies are also what we have extensively discussed in the third chapter (electronics, lasers, LED light, solar panels, imaging diagnostics, etc.), considered those of the first generation. However, when we talk about quantum technologies 2.0 today, we mainly refer especially to those that use the singular phenomena of quantum superposition and entanglement, as well as those technologies that are based on the manipulation of relatively complex quantum systems.

4.4.1 Quantum Computer

Just in the years when Aspect was doing his famous experiments, Feynman hypothesized that the physical world in its complexity could be deciphered through calculators that used quantum phenomena, namely quantum computers. In particular, he published in 1982 an article on the quantum computer in which he demonstrated that no classical calculator is able to simulate particular physical phenomena without an inevitable exponential slowdown in performance. On the contrary, a simulator or quantum computer would have been able to do it much more efficiently. The article of the great physicist was not ignored and in 1985 David Deutsch of Oxford University developed the first theory of quantum computability, inaugurating the new and fascinating field of quantum computation.

To understand the advantage of quantum computation compared to classical one, remember that the classical computer is based on bits that can only assume two states, for example high or low, true or false or simply zero or one (binary logic) and that are practically realized using electronic devices (transistors, diodes). So a bit of information is a two-state system and can also be encoded in a quantum system that provides two states such as, for example, photons with two different polarizations, two particles/atoms having two different spin states or a superconducting ring with two currents circulating in one direction or another. In this regard, quantum mechanics tells us that, in addition to the two basic logical states 0 and 1, a quantum bit (*qubit*) can be prepared in a coherent superposition of the two states analogous to the two states of Schrödinger's cat. This means that the qubit is simultaneously in the two states 0 and 1, and as a result, it can encode at a given moment both state 0 and state 1 and thanks to the phenomenon of entanglement it is possible to imagine two or more correlated qubits. We can, for example, think of representing a qubit with the spin of an electron, where the electron is not in a definite state, but in a superposition of states 0 and 1 corresponding to spin up and down or vice versa. A qubit can be described in quantum mechanical terms by the wave function $\Psi = a\Psi(0) + b\Psi(1)$ where $\Psi(0)$ and $\Psi(1)$ are respectively states 1 and 0, while a and b are coefficients whose square provides us with the probability of finding state 0 and 1 respectively after a measurement. Before the measurement the qubit can be simultaneously in the two states and therefore can encode both states.

Let's now consider a register of two entangled qubits *entangled*. A classical register so composed can represent, at any moment, only one number among the 4 possibilities; that is, the register can be found in only one of the 4 possible states 00, 01, 10 and 11, while a quantum register of 2 bits can represent at any moment all the possible states in a coherent superposition. If we consider 3 qubits the possible states increase to 8 and therefore a 3 qubit register can simultaneously encode 8 states. This results in a parallel process that greatly increases the computing capacity. If we add more qubits to the register, its ability to simultaneously represent states increases exponentially, in particular it increases by 2^N where N is the number of qubits. It is precisely the peculiar characteristic of being able to represent simultaneously all the available states that gives the quantum computer undeniable advantages over classical computers. It is therefore about the possibility of performing a perfect calculation in parallel. The American physicist-computer scientist Seth Lloyd argues: "Classical computation is like a solo, a pure melodic line that follows. Quantum computation is like a symphony: many melodic lines overlap each other".

However, reading the result of a quantum calculation is always a classical operation that inevitably causes the wave function to collapse, making the superposition of states disappear and from the quantum register only one number is always read, so only a part of the result of the process is usable. For these reasons, quantum algorithms focus more on the global properties of some functions rather than on individual numerical values.

A few years after David Deutsch's theory demonstrating the feasibility of quantum computing, the first quantum algorithms began to be developed. Let's briefly remember that in general an algorithm is a sequence of elementary operations or instructions that allow to solve a problem.

The first quantum algorithm was developed in 1992 by David Deutsch and Richard Jozsa, demonstrating for the first time the possibility of solving a computational problem with a speed significantly higher than that of an analogous classical algorithm. It is an algorithm of little practical interest but very important from a point of view of the feasibility of quantum computing. The algorithm provides a system of which the content is unknown (black box or oracle), the input of the system consists of a series of 2^N binary values and the output of the black box can only be 1 or 0. If the output is always 0 or 1, the function is defined as constant, but if the output is 1 for half of the inputs and 0 for the other half, the function is said to be balanced. The purpose of the aforementioned algorithm is to determine whether the function is constant or balanced. If after the first two binary inputs the black box provides two different results (0 and 1 or 1 and 0), it is immediately deduced that the function is balanced, but in the worst case it is necessary to query the oracle a number of times equal to half of the possible cases plus one, i.e., $(2^N/2) + 1 = 2^{(N-1)} + 1$.

The next quantum algorithms we will illustrate are of great practical interest. In 1994 Peter Shor demonstrated that the problem of factoring an integer could be efficiently solved on a quantum computer. Factoring a number means breaking it down into prime numbers (numbers divisible only by themselves), which multiplied together give us the original number. For example, the factorization of 35 is simply given by 7 and 5; in fact, 7 and 5 are two prime numbers whose product is 35. In this case the procedure is immediate, but if we think of 391, the operation is not immediate. If the number to be factored has more than 100 decimal digits, even supercomputers take a very long time. From a computational point of view, such a problem is said to have exponential difficulty, in the sense that the time and memory used increases exponentially with the increase in the digits of the number to be factored. This type of operation may seem of little interest, but it is actually the basis of the major encryption techniques currently used. Using Shor's quantum algorithm, the difficulty becomes polynomial, i.e., the computation time increases like a polynomial and not like an exponential. To realize how quickly the exponential trend increases compared to a polynomial one, consider for example n^2 (polynomial) and 2^n (exponential) for n = 20, in the first case we have $20^2 = 400$ in the second case $2^{20} = 1,048,576$. To be more concrete, if we consider a number with 2048 bits, equivalent to a number with 617 digits, its factorization could take several decades even using a supercomputer and only a few minutes with a quantum computer using Shor's algorithm. A similar thing happens for the Deutsch-Jozsa quantum algorithm.

Since most of the current cryptographic protocols are based precisely on the difficulty of factoring into prime numbers, Shor's algorithm has opened up new issues and concerns, highlighting that information security could be compromised by quantum cryptoanalysis. However, as we will see, the solution comes from the same quantum world from which Shor's algorithm was born.

In 1996 Lov Grover showed that the problem of searching in an unordered database can be sped up using quantum computation. To clarify, consider the classic example of the phone book: suppose we want to find a person's phone number of whom we know the name and surname; the operation is quite easy. Except for possible cases of homonymy, within a few seconds we manage to find the phone number. This is because the phone book is written in alphabetical order, so it is an ordered database if you look at it from the names of people. However, if we want to find the name and surname from the phone number, the operation is not simple at all. In fact, in this case it is an unordered database and to find the name corresponding to the phone number we know, we should start from the first page and flip through them all until we have found our number. This is therefore a serial search and no classic algorithm can improve the search method. The Groove quantum algorithm, on the other hand, allows for a search in parallel, greatly reducing search times. In other words, the aforementioned quantum algorithm allows us to read simultaneously more pages. This type of algorithm also undermines the security of some cryptographic codes and in particular of the *DES* (Data Encryption Standard) which requires a search among about 70 million billion possibilities, an operation that is almost impossible for a commercial classical computer but not for a quantum one that uses Grover's algorithm. In practice, the advantage is linked to the fact that, while a classic search algorithm in a disordered database of N elements must make at least N/2 steps, Grover's algorithm predicts a number of steps equal to the square root of N and therefore significantly fewer. If for example N = 1,000,000, the search with Grover's algorithm will be completed at most with 1000 steps, while 500,000 steps are necessary for a classic search algorithm, which is 500 times more in terms of computational times, this is a really significant difference.

To the mentioned algorithms, interesting and promising studies were added on the possibility of implementing quantum logic gates as well as programming languages for quantum computers (*Q-gol, qCGL, quantum C-Language*) aimed at developing a language that would allow the use of formalism similar to that of existing languages. We can therefore affirm that the results of quantum computing developed in the last 20 years of the last century were very flattering and left us hopeful.

At this point, only the physical realization of the first qubits was missing, that is, the quantum hardware was missing. We remember that classical bits are made with semiconductor elements, transistors that do not operate as amplifiers, but in saturation conditions, that is, their output is characterized by a state of zero voltage (state *0* of the bit) or by a state of finite voltage (state *1* of the bit). How are qubits made instead? How many types of qubits exist? It is not difficult to imagine that the realization of quantum bits is not simple and only the development of advanced technologies, like the one that has occurred in the first 20 years of the new millennium, has allowed their realization.

A qubit must first of all be a two-state system, that is, provide a state to which we can associate the state *0* and a second state to which to associate the state *1*. Nevertheless, that's not enough, it is necessary to be able to realize the superposition of states and maintain it for a reasonable time so that the qubit can operate, that is, the decoherence time must not be too short. We remember that decoherence is the process that destroys quantum superposition and as such represents the main problem for the realization of qubits and more generally of the quantum computer. Various prototypes of quantum bits have been realized, the most promising are those based on superconductors, trapped ions, and photons.

Superconducting qubits (Fig. 4.7), among the first to be realized, require cooling to a temperature close to absolute zero (0.01 K), have the advantage of being scalable and miniaturized like semiconductor chips and are controllable with electrical, magnetic or microwave signals. The two superimposed quantum states are generally represented by two energy states of the device or two electric currents that circulate in one direction or the other in a superconducting ring. The disadvantages are represented by the use of advanced

Fig. 4.7 Picture of superconducting quantum processor (Sycamore chip) used in Google's quantum computer. (Reproduced with permission from Springer Nature. Frank Arute et al., Nature, 574, 505, 2019)

cryogenic technologies to be able to reach extremely low temperatures and by fairly short decoherence times.

In trapped ion qubits, the use of appropriate electromagnetic fields allows the trapping and cooling of ions in a superposition of excited energy states. In particular, two orbitals of the atom are used to encode the states 1 and 0, and through lasers, the ions are excited, moving an electron from a lower energy orbital to a higher energy one. When an electron decays, it emits a photon, which is detected by special sensors that represent the eyes to see whether the atoms emit or do not emit photons and allow the encoding of the states 0 and 1 corresponding to the absence or presence of photon emission respectively. These qubits have the advantage of being very stable with reasonable decoherence times, but their realization requires the use of many lasers and therefore they do not have good scalability.

Certainly, the simplest quantum bits from a realization point of view are photonic ones, whose quantum states can correspond to two different polarizations of photons. They have fairly long decoherence times, but scalability is quite complicated, despite the possibility of implementing them on photonic chips.

There are also new very promising platforms such as neutral atoms trapped by focused laser beams, called *laser tweezers*. In these systems, to achieve the superposition of states, the energy levels of electrons or the spin states of atomic nuclei are exploited. Unlike trapped ions, with a single laser beam, appropriately divided into many laser tweezers, several hundred atoms can also be controlled and potentially even a few thousand. In addition, the neutral atoms, a few micrometers apart from each other, show a very long decoherence time (several seconds).

Another very interesting technique, but still in its infancy, involves the realization of qubits within a semiconductor, using electric fields to encode quantum information in the spins of individual electrons. The great advantage of these semiconductor-based qubits would be to use the consolidated semiconductor technology capable of creating reliable devices with nanometric resolution.

As seen, there are several types of quite promising qubits, but at the moment it is not easy to predict which qubit will prevail in the race towards the definitive quantum computer.

Naturally, in addition to the difficulty of creating reliable qubits, the other fundamental aspect for the realization of a quantum computer is to connect the qubits together in ways to create a network of qubits that are quantumly correlated (*entangled*) with each other. A single qubit is not very useful: to be able to perform calculations exploiting the power of perfect quantum parallel computing, at least a few dozen qubits are needed. The first quantum coupling between two qubits was realized in 1995 by Ignacio Cirac and Peter Zoller, using trapped ion qubits. They created the first prototype of a quantum logic gate, the *controlled-not* (CNOT) gate, a fundamental component for the implementation of the quantum computer. A CNOT gate performs the following function: given a binary input (*0* and *1*) the CNOT inverts the second bit only if the first bit is *1*.

Since the realization of the first prototypes of qubits and logic gates, extraordinary progress has been made leading to the creation of the first quantum computers (Fig. 4.8), also thanks to the interest of large multinationals such as Google, IBM and Intel and the birth of innovative start-up companies like the Canadian D-Wave, Xanadu, the American IonQ, the Rigetti, the Dutch QuantWare and the Italian-Anglo-American SEEQC.

In particular, in 2019 Google created a quantum computer with a processor (Sycamore) of 53 qubits based on superconducting devices (Fig. 4.7) and demonstrated the so-called *quantum supremacy* that is, the ability to solve in a few seconds or minutes a problem that would have taken about 10,000 years with the current supercomputers. In June 2022, the Canadian Xanadu presented a quantum computer made in collaboration with the *National Institute of Standards and Technology*, USA, with 216 photonic qubits capable of performing a complex calculation in just 36 millionths of a second and that would have taken more than 9000 years if performed with a classic supercomputer.

Fig. 4.8 Photo of the internal parts of a quantum computer made by IBM based on superconducting qubits. The cables used to control the qubits are evident. The operation of such a computer requires temperatures close to absolute zero implying sophisticated cryogenic techniques. (Reproduced with permission from Springer Nature. Philip Ball, Nature, vol. 599, 542, 2021)

In November 2022, IBM introduced a new quantum processor (Osprey) with 433 superconducting qubits, while in december 2023 one with 1121 qubits (Condor) was presented. IonQ, on the other hand, is focusing on qubits made with single trapped ions, planning to replace the ytterbium atoms used so far with those of barium. These offer greater advantages compared to the previous ones in terms of a lower error rate, greater scalability, and better quantum state detection.

In some cases quantum supremacy has been refuted, in the sense that it has been shown that the same problem could be solved with a supercomputer in a reasonable time as happened in 2019 with Google's quantum computer. Indeed, shortly after, IBM scientists demonstrated that the same problem solved by Google's quantum computer could be solved in just over 2 days instead of the 10,000 years that Google scientists were talking about.

Today there is a tendency not to talk about quantum supremacy but rather about *quantum advantage*, also because due to its peculiarities the quantum computer is not usable to solve any computational problem and those problems where there is an actual quantum supremacy, at the moment are only demonstrative and of very limited applicative interest. The realization of the quantum advantage, understood as optimization and improvement of existing processes, has a much more pragmatic approach, and aims at solving specific problems using the integration between classic and quantum algorithms.

Quantum computation will be able to be used to study complex systems with prospects of important discoveries and further technological progress in many fields such as: medicine, biology, chemistry, pharmacology, bioengineering, atmospheric physics, artificial intelligence, transport etc. Think for example of the advantage of doing genetic mappings in a very short time or of understanding at the atomic level the functioning of biomolecules or also the possibility of doing complex simulations of environmental and atmospheric events.

4.4.2 Quantum Cryptography and Teleportation

In addition to the quantum computer, another application of the quantum phenomena exposed in this chapter is the *quantum cryptography*. In general, when we talk about cryptography we refer to the techniques that allow to transmit messages in full secrecy preventing unauthorized people from reading or modifying the messages.

With the progressive digitization, it is absolutely necessary to use digital cryptographic techniques capable of protecting our sensitive data. Consider bank transactions that today take place almost entirely through home banking, or the communication of sensitive messages via email or even the purchase with credit cards on websites.

Historically, to prevent third parties from reading messages, encrypted writings were used using appropriate codes known only to the sender and the recipient. The message was delivered by hand or, in more recent times, transmitted telegraphically.

A very simple example of encryption is that of Caesar, one of the oldest known ciphers. First described by the Roman biographer Svetonio, the aforementioned cipher was frequently used by Caesar for his private correspondence. It is a *substitution* cipher whose operating principle can be easily understood with the following example: suppose we want to encrypt the name MARIO using the algorithm that replaces each letter with one that is three places away in the Latin alphabet. M becomes P, A becomes D and so on until we get instead of MARIO the word PDULR which has nothing to do with the word we started from. If you do not know the encryption code, it is very difficult to trace back to the word. With the advent of computers and the internet, the basic principle has remained the same but the transmission mode has changed. Messages continue to be encrypted and to decrypt them it is necessary to have a computer key, or an alphanumeric string, which allows through a mathematical algorithm to encrypt and decrypt the message. If the key is unique, it is called *symmetric cryptography* and has the disadvantage of using a large number of keys, one for each pair of interlocutors, since the key must be secret. If instead the key for encryption is different from that of decryption, it is called *asymmetric cryptography* which is the most used. The key for encryption is public, everyone can see it, that of decryption is instead secret and only the recipient holds it.

The name key evokes a box with a lock. To better understand the operating principle of symmetric protocols with secret key and asymmetric with public key, we can imagine just a mechanical combination box. In the case of symmetric cryptography, we place our secret message inside the box and close it, then we send the box to the recipient who can read the message after opening the box with the opening code received from the sender through a secure channel (key sending). In the case of asymmetric protocol, the recipient of the message sends the open box to the sender, of which only he knows the opening code. The sender after putting the message inside, closes it and sends it back to the recipient who can easily open it since he has the opening code (no key sending).

The only perfect security classical cryptography algorithm is the Vernam cipher, developed by Gilbert Vernam in 1917. It involves the use of a single key (symmetric cryptography) with a size equal to the message to be transmitted and usable only once. For this reason, the Vernam cipher is also called *One Time Pad* (OTP), which literally means disposable notebook/cipher. The big disadvantage of this cryptographic system is the need to transmit keys through a secure channel, which greatly limits its use.

Instead, the most widely used cryptography is asymmetric based on the RSA algorithm, named after the three researchers who developed it in 1977, Ron Rivest, Adi Shamir and Leonard Adleman. The aforementioned algorithm uses the factorization into prime numbers of an integer with many digits (even more than 100 decimal digits). The public key is linked, through a known algorithm, to the integer, while the secret key is linked to the prime numbers whose product is equal to the public key number. Therefore, to seize the secret key, it is necessary to factorize the public number into prime numbers. Cryptography based on the RSA protocol is quite secure. Even using supercomputers, the time needed to determine the secret key is so great that it is unlikely that a potential intruder will be able to decrypt the message and gain access to sensitive information.

However, the potential of the quantum computer, particularly with regard to the ability to factorize prime numbers in reasonable times using Shor's algorithm, could seriously jeopardize cybersecurity. Nevertheless, it is precisely the principles on which the quantum computer is based that avert this possibility. In fact, the superposition and entanglement of quantum states have allowed the birth of quantum cryptography, which allows the creation of practically inviolable keys.

The first quantum cryptography protocol was developed by Charles Bennett and Gilles Brassard in 1984 and is named BB84 after the initials of its authors and the year it was developed. This protocol is based on the use of single photons that are polarized through appropriate filters along 4 directions: vertical, horizontal, oblique at 45° and 135°. Photons with vertical and 45° oblique polarization are generally assigned the value *1* and those with horizontal and 135° oblique polarization the value *0* (Fig. 4.9).

The secret key is sent to the recipient through a sequence of single photons with different polarizations and generally traveling on optical fibers. The photons, after passing through additional filters randomly arranged by the recipient, are detected by special detectors capable of detecting even a single photon and determining the polarization. The filters used by the recipient are orthogonal, either horizontal/vertical or oblique, either at 45° or 135°.

According to the laws of quantum mechanics, it is impossible to observe the state of the photons without modifying them. Therefore, not knowing a priori the state of polarization of the photons, when using a filter different from that of the sender, an inevitable modification of the state of polarization

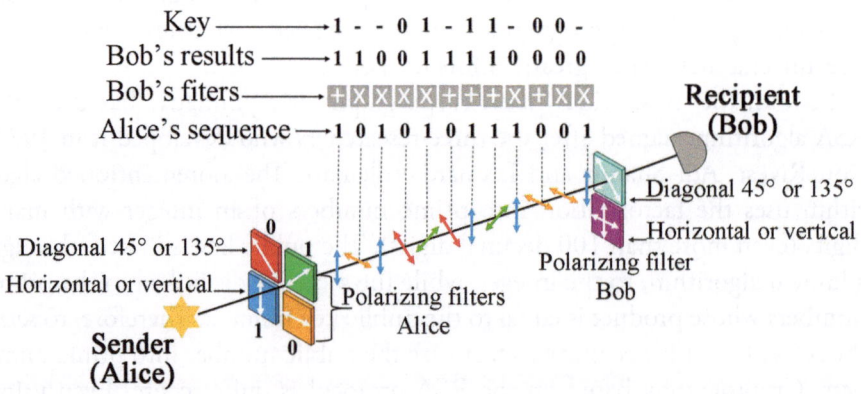

Fig. 4.9 Scheme of the BB84 quantum cryptography protocol based on the use of polarized single photons. A potential intruder in determining the polarization of the photons sent by Alice, will inevitably make mistakes. Therefore, the intruder will send to Bob a key that is certainly different from Alice's and since before starting the transmission Alice and Bob compare a part of the key, in case there has been an actual intrusion they will notice it

of the photon is introduced. For example, if the sender (Alice) sends a photon with vertical polarization and the recipient (Bob) uses an orthogonal filter, the photon will pass preserving its polarization, if instead he uses an oblique filter, a photon polarized at 45° or one at 135° is obtained with a 50% probability since a vertical or horizontal photon is in a superposition of the two states at 45° and 135°. To determine the exact polarization of the photons Bob can use additional filters: in case he has placed an orthogonal filter, he can use a vertical filter and observe whether the photon passes or not. If it passes, it is certainly a photon with vertical polarization to which to assign the value *1*, in the case where it does not pass, the photon is certainly of the horizontal type to which the value zero is attributed. A similar procedure can be carried out to determine the polarization at 45° or 135° if an oblique filter has been used. Obviously, if the photon sent by Alice is of the vertical or horizontal type (oblique at 45° or 135°) and Bob uses an orthogonal (oblique) filter, he will be able to determine exactly the polarization of the starting photon. Otherwise, the probability is 50%. Therefore, Bob will make a series of errors in receiving the message, which can be estimated at a percentage equal to 25%, that is, the sequence of *1* and *0* will contain on average a quarter of incorrect values. To eliminate them, Bob contacts Alice, even on an insecure channel, and communicates to her the sequence of filters used; at this point Alice indicates to Bob the wrong filters but not the polarization of the photons. With this information, the recipient can eliminate the non-matching values and the remaining sequence matches that of the sender. The sequence thus obtained, smaller than the one sent, represents the raw or secret key.

What happens if there is an intruder who intervenes between Alice and Bob trying to intercept the sequence sent by the sender? Obviously, he too will make mistakes and will transmit to Bob a sequence different from that transmitted by Alice. The probability that the intruder has to reconstruct the exact sequence sent by Alice is $(3/4)^n$, where n is the number of photons. For $n = 50$, the probability is less than one in a million; therefore, the intruder will send a key to Bob that is certainly different from Alice's. Before starting the transmission, Alice and Bob compare a part or more parts of the key, for example the first ten numbers and/or the last ten. If these sub-sequences coincide, they can be sure that there has been no interference and start the transmission of the message; otherwise in case of differences even of a single number, they can repeat the operation from scratch or find alternative solutions. Naturally, the parts of the key transmitted on the insecure channel are not used in the final secret key.

Unlike the classic RSA system, the BB84 protocol only allows the keys for decoding the message to be exchanged securely, which can be sent using a very secure cipher like Vernam's. For this reason, when talking about quantum cryptography, the term *Quantum Key Distribution* (QKD) is often used.

Another important quantum cryptography protocol was developed by Artur Ekert in 1991 (E91), and is based on quantum *entanglement*. In this case, a source emits particles or photons *entangled* with opposite spins and directed to the two interlocutors who perform spin or polarization measurements. The procedure for generating the secret key is very similar to that used for the BB84 protocol. In this case, the inviolability of the key is ensured by the entanglement of the particles. In fact, if an intruder tried to read the state of the photons/particles, he would destroy the quantum correlations, resulting in a different measurement of the spin or polarization made by the two interlocutors (Alice and Bob), who would deduce the presence of the intruder.

The first currency transaction protected by quantum cryptography was carried out in 2004 between two Austrian banks; in the same year, both in Cambridge (England) and in Boston (USA), the first quantum cryptography network on optical fibers based on the BB84 protocol, known as *DARPA Quantum Network*, was implemented, which operated continuously for 3 years.

Using the *Micius* satellite and fiber connections, in 2017 China managed to conduct a 75-min video conference protected by quantum cryptography, between the Chinese city of Xisinglong (about 300 km from Beijing) and the Austrian city of Graz, which are about 7600 km apart, demonstrating the possibility of creating a global quantum network.

On the occasion of the G20 in Trieste in 2021, the first inter-European quantum network was tested, also based on the BB84 protocol and connecting Italy, Slovenia, and Croatia. The network was created thanks to the collaboration of the Italian National Research Council (CNR), the University of Florence (Italy), and the University of Denmark with the support of various companies.

At the beginning of 2024, the metropolitan quantum communication network of Napoli (Italy) was inaugurated, connecting the CNR research area of Pozzuoli with the Federico II University Campus of San Giovanni a Teduccio at the Meditech Competence Center and the Leonardo Laboratories in Pomigliano. The aforementioned network is based on a completely Italian technology, developed by the collaboration between CNR, the National Institute of Metrological Research, the University of Naples, and various Italian companies (Quantum Telecommunications Italy, TIM, ThinkQuantum, Cisco, Exprivia).

As cybersecurity is a strategic sector, many large companies, such as IBM, Toshiba, and TIM, are investing in these cutting-edge communication technologies.

One of the problems with quantum cryptography, when considering large distances, is the attenuation of the signal, which is impossible to amplify without destroying the quantum information.

A solution to this problem could come from *quantum teleportation,* first proposed in 1993 by Charles Bennett and his research group. Contrary to what the resounding name would suggest, quantum teleportation has nothing to do with that of fiction and in particular the famous television series *Star Trek.* In the case of quantum teleportation, there is no transfer of matter, also because it is physically prevented by the Heisenberg uncertainty principle: the information to be transported for the reconstruction of the matter that one would like to teleport would mostly be destroyed during the measurement phase, making materialization impossible. The purpose of quantum teleportation is to transfer a state from one point to another using once again the phenomenon of *entanglement.*

Suppose Alice wants to transfer the state of a photon X to Bob. She performs the following procedure, which we report in an extremely simplified version: she uses a pair of *entangled* photons that reach her and Bob; the state of the photon X to be transferred to Bob is made to interact with the photon A of the pair received by Alice who makes a measurement on the X-A pair. Due to entanglement, the measurement made by Alice instantly affects the state of Bob's photon B. At this point, Alice calls Bob via a classical channel (phone, email) to communicate the outcome of her measurement and Bob, with appropriate operations, subtracts the effect of the pair of *entangled* photons and transforms the state of the photon B into one identical to the photon X. In other words, an identical copy of a state is reproduced even at very great distances without any transfer. Note that there is no violation of the principles of Special Relativity, as the process is completed with a classical communication channel.

Today, quantum teleportation is a well-established reality. From the first experiments carried out on photons in 1997–1998, we have moved on to teleportation of atomic states in 2004, leading up to the teleportation of single photons over long distances (1400 km) via satellites in 2017.

4.4.3 Other Quantum Technologies

New quantum technologies do not only refer to computers, cryptography, and communications, but also to other sectors such as quantum sensors, *quantum imaging* and *quantum simulation* through cold atoms.

Quantum sensors allow to obtain extraordinary sensitivities limited only by the principles of quantum physics. They are typically used to measure weak magnetic fields, electrical signals, frequency, time, gravitational field, but also the magnetic moment (spin) of single elementary particles and small displacements. To avoid going into too technical details, we will not illustrate the details of the operating principles of quantum sensors but we will only mention the main quantum sensors and their applications.

In the third chapter, we have already seen some examples of quantum sensors, the SQUID, which by exploiting the tunnel of electron pairs and wave function interference, can measure magnetic fields so small that they can even detect the very weak magnetic fields produced by neuronal currents within our brain. However, there are other types of quantum magnetic sensors, such as atomic ones based on the interaction between the magnetic field to be measured and the electronic spins of an alkali atom vapor (sodium, potassium). Having a very high magnetic field sensitivity and not requiring cryogenic liquids for their operation, these atomic sensors could in the future replace the established SQUID sensors which, being made of superconductors, require the use of helium or liquid nitrogen. Then there are sensors based on the *vacancies* of nitrogen atoms in diamonds, particularly interesting for their sensitivity in measuring magnetic moments of magnetic nanoparticles or single molecules but can also be used for temperature, pressure, and very small rotation measurements. Particularly interesting quantum sensors are also those based on ultra-cold atoms for ultra-sensitive measurements of time, frequency, and gravity acceleration or atomic cell sensors successfully used to realize high-precision gyroscopes.

Finally, particularly important for the same quantum technologies such as photonic quantum computers and quantum cryptography, are single photon detectors. These extraordinary sensors are largely based on the photoelectric effect, but the most performing ones use superconducting nanowires or optical cavities.

Another very interesting example of quantum technology is *quantum imaging*, which allows you to see even objects illuminated so weakly as to seem invisible. When we take a photograph, we know well that, in case of poor lighting, the photo is blurred and not very sharp. This is essentially due to the fact that the number of photons reaching the camera sensor is so small and fluctuating that it generates a noise (*shot noise*) responsible for the poor resolution and sharpness of the image. By exploiting the quantum correlation of photons it is possible to acquire good images even in the presence of this optical noise. A technology known as *quantum plenoptic imaging* is used: the few photons coming from the subject are divided into two beams by a suitable lens and sent to two sensors that reproduce the in-focus images of two different planes of the three-dimensional scene that you want to acquire. Essentially, images with different perspectives are acquired simultaneously. The measurement of the spatial and temporal correlation of the two beams and the relative processing allow to reconstruct the three-dimensional scene without losing resolution and depth of field. Naturally, this technology is not limited to photographs but can be applied to any device that captures images such as microscopes, cameras, satellites, and biomedical instruments.

Of great interest is quantum simulation using ultra-cold atoms. This quantum technology allows the simulation of very complex systems using atoms cooled to a temperature close to absolute zero. As seen in the previous chapter, when we discussed Bose-Einstein condensation, at such low temperatures the atoms are practically immobile and through appropriate manipulation can assume precise geometric positions as happens for nuclei within crystalline solids. It is therefore possible to reconstruct a physical system very similar to the one you want to study with the difference that it can be controlled and manipulated at will. Quantum simulation based on ultra-cold atoms is used to study magnetic materials and superconductors but also to simulate and improve the performance of certain types of lasers such as the quantum cascade laser in which knowledge of electron dynamics is fundamental to optimize laser performance.

Although we will not discuss them, quantum metrology and new quantum materials that offer interesting perspectives from an application point of view should certainly also be mentioned.

4.4.4 The Strategic Sector of Quantum Technologies

As seen, the new quantum technologies concern, in addition to the quantum computer, other strategic fields such as cryptography, communications, internet, and advanced sensing. For this reason, quantum technologies are already considered a strategic sector to invest in and will most likely become as important in the future as the semiconductor field for which there are clear economic and geopolitical interests. Just think of the strategic importance of the two major semiconductor technology centers, Silicon Valley in the United States and Taiwan.

On quantum technologies, in addition to large companies, large amounts of money are also invested by public research in all industrialized countries, especially by China and the USA, which have recently invested tens of billions of dollars in this strategic sector. The European Union launched a flagship project on quantum technologies in 2018 with funding of about 1 billion euros. In 2022, Italy, as part of the national recovery and resilience plan, funded a project of about 320 million euros for the creation of a state-of-the-art national digital infrastructure dedicated to big data processing and quantum computation and a project of about 120 million euros for the establishment of an Italian consortium that will carry out competitive and innovative research in the field of quantum sciences and technologies. Both projects involve the participation of Universities, public research bodies, and private companies.

The last question we ask ourselves: when will the new quantum technologies be available on the market? The quantum computer is potentially an extraordinary tool and certainly represents a significant leap in the field of computation, but it is still difficult to predict how much time is needed to have common or even commercial quantum calculators. There are still several problems to solve such as decoherence that destroys correlated states and makes a quantum computer equal to a classic one. Error correction approaches due to decoherence are certainly effective, but currently the probability of reading the correct results at the end of a quantum operation is not yet satisfactory. Instead, other quantum technologies such as cryptography, sensors, and imaging have already demonstrated their operation and effectiveness, suggesting a commercialization in a short time.

Indeed, Feynman had a good intuition, in fact, starting from Aspect's pioneering experiment, the second quantum revolution began, starting from phenomena such as entanglement and quantum coherence, leading to the development of new quantum technologies, dubbed quantum technologies 2.0.

In conclusion, we can affirm that in physics and more generally in science, no thought, conjecture, or idea should be overlooked or dismissed superficially, as in many cases, behind an apparently uninteresting idea, true revolutions are hidden. When in 1946, Felix Bloch and Edward Purcell discovered nuclear magnetic resonance, they could never have imagined that one day that phenomenon would change the world of diagnostic imaging, or when in 1915 Einstein presented his theory of General Relativity he would never have thought that one day, thanks to relativistic corrections, GPS would guide us to our destinations with an accuracy of a few meters or even the discovery of the transistor effect in 1947 would never have imagined the birth of semiconductor electronics, which can be considered, without exaggeration, the most important technological revolution of the twentieth century.

The new quantum technologies were born from philosophical speculations on the conceptual foundations of quantum mechanics and no one would ever have thought that from those philosophical and at times metaphysical debates, a second quantum revolution would be born, most likely destined to change our way of life once again.

With good probability we can affirm that the twenty-first century will be characterized by quantum technologies 2.0 and artificial intelligence which in recent years has made giant strides. Unlike the latter which tends to replace man, quantum technologies will most likely be at the service of man as tools to understand and solve complex problems, to protect and transmit data securely, to acquire increasingly clear and resolved images, to measure physical quantities and reveal objects with unprecedented precision and many other things that the so-called second quantum revolution is allowing us to develop. However, it is not excluded that the potential of quantum computing can make artificial intelligence even more efficient and in some ways more disturbing, but we hope that the good sense of *homo sapiens* will be able to control any side effects of the powerful technologies we will have to deal with in the future.

Recommended Readings[1]

Theory of Special and General Relativity

Cowen R. (2019), Gravity's Century: From Einstein's Eclipse to Images of Black Holes, Harvard University Press (ISBN 0674974964).

Einstein A. (1966, first published 1922), The Meaning of Relativity, Princeton University Press (ISBN 9780691023526).

Ferreira P.G. (2014), The Perfect Theory: A Century of Geniuses and the Battle over General Relativity, Houghton Mifflin Harcourt (ISBN 9780547554891).

Gardner M. and Ravielli A. (1997), Relativity Simply Explained, Dover Publications (ISBN 9780486293158).

Greene B. (2005), The Fabric of the Cosmos: Space, Time and the Texture of Reality, Penguin (ISBN 9780141011110).

Hartle J.B. (2021), Gravity: an Introduction to Einstein's General Relativity, Cambridge University Press (ISBN 1316517543).

Hawking S. (2011), *A Brief History Of Time: From Big Bang To Black Holes*, Transworld Publishers Ltd (EAN: 9780857501004).

Kaku M. (2004), Einstein's Cosmos: How Albert Einstein's Vision Transformed Our Understanding of Space and Time, W.W. Norton (ISBN 9780393327007).

Pais A. (2005), *Subtle Is the Lord: The Science and the Life of Albert Einstein,* Oxford University Press (ISBN 9780192806727).

Rovelli C. (2016), *Seven Brief Lessons on Physics*, Riverhead Books (ISBN 0399184414).

Rovelli C. (2023), *White Holes*, Riverhead Books (ISBN 0593545443).

---

[1] Below are some popular texts on the topics covered in the book.

© The Editor(s) (if applicable) and The Author(s), under exclusive license to Springer Nature Switzerland AG 2025
C. Granata, *A Journey into Modern Physics*, https://doi.org/10.1007/978-3-031-77775-2

Schmitz W. (2022), Understanding Relativity. A Conceptual Journey Into Spacetime, Black Holes and Gravitational Waves, Springer Nature (ISBN 9783031172182).

Stannard R. (2008), Relativity: A Very Short Introduction, Oxford University Press (ISBN 9780199236220).

Strohm T. (2023), Special Relativity for the Enthusiast, Springer Nature (ISBN 9783031219238).

Wilkinson Carl. (2020), Albert Einstein's Theory of Relativity: Words That Changed the World Laurence King Publishing (ISBN 1786277506).

Quantum Physics

Aczel A. D. (2002), *Entanglement. The greatest mystery of physics*, John Wiley & Sons Ltd, 2004 (ISBN 0470850477).

Al-Khalili J. (2020), *The World According to Physics*, Princeton University Press (EAN 9780691182308).

Al-Khalili J. (2004), *Quantum: A Guide for the Perplexed*, Orion Publishing (ISBN 1841882380).

Aspect A. (2024), Einstein and the quantum revolutions, 0226832015 (ISBN 0226832015).

Baggott J. (2013), *Higgs. The invention and discovery of the 'God Particle'*, Oxford University Press (ISBN 0199679576).

Bell J. S. (2004), *Speakable and unspeakable in quantum mechanics*, Cambridge University Press (EAN 9780521523387).

Bricmont J. (2017), Quantum Sense and Nonsense, Springer International Publishing (ISBN 9783319652702).

Dyakonov M. I. (2020), Will We Ever Have a Quantum Computer?, Springer Nature Switzerland (ISBN 9783030420185).

Eredidato A. (2020), *Ever Smaller: Nature's Elementary Particles, From the Atom to the Neutrino and Beyond*, The MIT Press (ISBN: 0262043866).

Feynman R. P. (1990), *QED. The strange theory of light and matter*, Penguin (ISBN 9780140125054).

Feynman R. P. (1992), *The Character of Physical Law*, Penguin (ISBN 0140175059).

Feynman R. P. (1994), *Six Easy Pieces: Essentials of Physics Explained by Its Most Brilliant Teacher*, Perseus Books (ISBN 0201409569).

Gato-Rivera B. (2021), *Antimatter*, Springer Nature Switzerland, (ISBN 9783030677909).

Gilmore R. (1995), *Alice in Quantumland: An Allegory of Quantum Physics*, Copernicus (ISBN 0387914951).

Gribbin J. (1985), In Search Of Schrodinger's Cat: *Quantum Physics And Reality*, Black Swan (ISBN 9780552125550).

Jaeger L. (2018), The Second Quantum Revolution. From Entanglement to Quantum Computing and Other Super-Technologies, Springer Nature Switzerland, (ISBN 9783319988238).

Kane G. L. (1995), *The garden of particles. How and why the physics of particles is changing our way of conceiving the Universe*, Perseus Books (ISBN 0201407809).

Kembhavi A. and Khare P. (2020), Gravitational Waves. A New Window to the Universe, Springer Nature Singapore Pte Ltd (ISBN 9789811557088).

Kumar M. (2007), Quantum: Einstein, Bohr and the Great Debate About the Nature of Reality, Icon Books (ISBN: 9781848310353).

Lederman L. M. and Christopher H. T. (2011), *Quantum Physics for Poets*, Prometheus Books (ISBN: 1616142332).

Lederman L. M. and Teresi D. (1993), *The God Particle*, Houghton Mifflin Harcourt (ISBN 0395558492).

McFadden J. and Al-Khalili J. (2015), *Life on the Edge: The Coming of Age of Quantum Biology*, Crown Publishing Group (NY) (ISBN 0307986810).

Rovelli C. (2022), *Helgoland: The Strange and Beautiful Story of Quantum Physics*, Penguin Books Ltd (EAN 9780141993270).

Schrödinger E. (2009), *My View of the World*, Turin, Cambridge University Press (ISBN 0521090482).

Silvestrini P. (2023), *Synchronous Physics*. Non-locality in the science of the millennium, Youcanprint (ISBN 9791221432664).

Tonelli G. (2024), *Matter. The Magnificent Illusion*, Polity Press (ISBN: 1509564144).